Marc Tänzer

Epigenetic biomarkers in colorectal cancer

Marc Tänzer

Epigenetic biomarkers in colorectal cancer

Novel epigenetic biomarkers for diagnosis, prognosis and response prediction in colorectal cancer

Südwestdeutscher Verlag für Hochschulschriften

Impressum/Imprint (nur für Deutschland/only for Germany)
Bibliografische Information der Deutschen Nationalbibliothek: Die Deutsche Nationalbibliothek verzeichnet diese Publikation in der Deutschen Nationalbibliografie; detaillierte bibliografische Daten sind im Internet über http://dnb.d-nb.de abrufbar.
Alle in diesem Buch genannten Marken und Produktnamen unterliegen warenzeichen-, marken- oder patentrechtlichem Schutz bzw. sind Warenzeichen oder eingetragene Warenzeichen der jeweiligen Inhaber. Die Wiedergabe von Marken, Produktnamen, Gebrauchsnamen, Handelsnamen, Warenbezeichnungen u.s.w. in diesem Werk berechtigt auch ohne besondere Kennzeichnung nicht zu der Annahme, dass solche Namen im Sinne der Warenzeichen- und Markenschutzgesetzgebung als frei zu betrachten wären und daher von jedermann benutzt werden dürften.

Coverbild: www.ingimage.com

Verlag: Südwestdeutscher Verlag für Hochschulschriften GmbH & Co. KG
Heinrich-Böcking-Str. 6-8, 66121 Saarbrücken, Deutschland
Telefon +49 681 37 20 271-1, Telefax +49 681 37 20 271-0
Email: info@svh-verlag.de

Approved by: München, TU, Diss., 2012

Herstellung in Deutschland:
Schaltungsdienst Lange o.H.G., Berlin
Books on Demand GmbH, Norderstedt
Reha GmbH, Saarbrücken
Amazon Distribution GmbH, Leipzig
ISBN: 978-3-8381-1464-4

Imprint (only for USA, GB)
Bibliographic information published by the Deutsche Nationalbibliothek: The Deutsche Nationalbibliothek lists this publication in the Deutsche Nationalbibliografie; detailed bibliographic data are available in the Internet at http://dnb.d-nb.de.
Any brand names and product names mentioned in this book are subject to trademark, brand or patent protection and are trademarks or registered trademarks of their respective holders. The use of brand names, product names, common names, trade names, product descriptions etc. even without a particular marking in this works is in no way to be construed to mean that such names may be regarded as unrestricted in respect of trademark and brand protection legislation and could thus be used by anyone.

Cover image: www.ingimage.com

Publisher: Südwestdeutscher Verlag für Hochschulschriften GmbH & Co. KG
Heinrich-Böcking-Str. 6-8, 66121 Saarbrücken, Germany
Phone +49 681 37 20 271-1, Fax +49 681 37 20 271-0
Email: info@svh-verlag.de

Printed in the U.S.A.
Printed in the U.K. by (see last page)
ISBN: 978-3-8381-1464-4

Copyright © 2012 by the author and Südwestdeutscher Verlag für Hochschulschriften GmbH & Co. KG and licensors
All rights reserved. Saarbrücken 2012

Zusammenfassung

Das kolorektale Karzinom ist in Deutschland bei Männer und Frauen die zweithäufigste Krebserkrankung. Die Heilungschancen durch Operation und Chemotherapie hängen entscheidend vom Krankheitsstadium ab. In dieser Arbeit wurden molekulare Marker zur Verbesserung der Diagnostik, Prognose und Responseprädiktion für das kolorektale Karzinom hinsichtlich ihrer Funktion im Tumor funktionell untersucht. Drei epigenetisch regulierte Gene (TFAP2E, TUSC3, RXFP3) wurden durch Transfektion in kolorektale Karzinomzelllinien (jene ohne endogene Expression dieser Gene) funktionell mittels Zellkulturassays auf Veränderungen hinsichtlich Proliferation, Apoptosis, Migration und Adhäsion untersucht, da dies die wesentlichen Kennzeichen von Krebszellen darstellen und sich somit die Rolle der Kandidatengene in diesem Kontext herausarbeiten ließ. Ein untersuchter Marker -RXFP3- eignet sich am besten für diagnostische Zwecke, ein weiterer -TUSC3 - als prognostischer Faktor und TFAP2E stellte sich als Responseprädiktor für die Therapie mit einem Standardchemotherapeutikum heraus.

Synopsis

This work describes the functional analysis of three epigenetic regulated genes (TFAP2E, TUSC3, and RXFP3) as biomarkers for diagnosis, prognosis and response prediction of colorectal cancer. From a panel of 12 candidate marker genes, these three were selected according to their expression and promoter hypermethylation in colorectal cancer cell lines and tumor samples from colorectal cancer patients as well as other criteria (literature, known functions, clinical characteristics). The coding sequences of all three marker genes were then transfected into colorectal cancer cell lines which lacked endogenous expression of these genes and the transfected cells were then functionally analysed using cell culture based assays. Changes in cell proliferation, apoptosis, migration and adhesion, stress response as well as response to chemotherapeutic agents were tested as those are hallmarks of cancer cells and could therefore be used to elucidate the role of the selected three marker genes in the tumor cells. Microarray based expression analysis was used for identification of potential downstream interaction factors and validated via quantitative PCR, chromatin immunoprecipitation and luciferase reporter assays. In summary, RXFP3 performed best for diagnostic purposes, TUSC3 could be useful as a prognostic tool since its hypermethylation correlates with lymph node invasion and TFAP2E could predict patient response to treatment with a standard chemotherapeutic agent.

Table of Contents

1. Introduction 1
   1.1 Colorectal Cancer - Statistics 1
   1.2 Biology 2
      1.2.1 Molecular changes 4
      1.2.2 Risk Factors 5
      1.2.3 Diagnosis 6
      1.2.4 Staging 8
      1.2.5 Treatment 10
   1.3 Epigenetics and Epigenomics 14
      1.3.1 Histone modifications and the histone code 15
      1.3.2 DNA Methylation 17
      1.3.3 Altered DNA Methylation in Cancer 24
      1.3.4 Histone modifications in cancer 29
      1.3.5 Epigenetic therapy in cancer 31
      1.3.6 Epigenetic biomarkers for cancer 39
2. Methods 42
   2.1 Patients and tissue samples 42
   2.2 DNA and RNA extraction 42
   2.3 Bisulfite Treatment 43
   2.4 Primer design 44
   2.5 MethyLight/MS-HRM analysis 46
   2.6 Reverse transcription polymerase chain reaction (RT-PCR) 51
   2.7 Cell culture and 5-aza-cytidine treatment 52
   2.8 Reporter and expression vectors and subcloning 53
   2.9 Generating of clones with stable overexpression 54
   2.10 Immunofluorescence and Immunoblotting 55
   2.11 Transient transfections and luciferase assays 57
   2.12 Chromatin immunoprecipitation 57
   2.13 Expression Microarray and verification of target candidates 58
   2.14 Stress resistance and cell survival assays after drug exposure 59
   2.15 Invasion and Adhesion 60
   2.16 Colony Formation 60
   2.17 Statistics 60
3. Goal and Purpose of this Thesis 62
4. Results 63
   4.1 Markers from Epigenomics 63
   4.2 Marker selection 65
   4.3 Marker Evaluation 70
   4.4 Functional Analysis 77
      4.4.1 Cell clones 77
      4.4.2 Microarray 77
      4.4.3 Cell Assays - TFPA2E 81
      4.4.4 Cell Assays - TUSC3 86
      4.4.5 Cell Assays RXFP3 88
   4.5 Marker Validation (TFAP2E) 88
5. Discussion 91
   5.1 General 91
      5.1.1 Markers TFAP2E, TUSC3, RXFP3 91
6. References 99
7. Tables, Figures, Abbreviations 112
8. Appendices 122

1. Introduction

1.1 Colorectal Cancer - Statistics

Globally, colorectal cancer is the third most commonly diagnosed cancer in males and the second in females, with over 1.2 million new cancer cases and 608,700 deaths estimated to have occurred in 2008.[1] While colorectal cancer death rates have been decreasing in several Western countries, rates continue to increase in many countries.[2] Colorectal cancer incidence rates are rapidly increasing in several areas historically at low risk, particularly in Central and South America and Eastern and Southern Europe. Such unfavorable trends are thought to reflect a combination of factors including changes in dietary patterns, obesity, and an increased prevalence of smoking.[3] Since the number of global cancer deaths is projected to increase over the next two decades, this number is expected to increase as well. In western countries alone, colorectal cancer is the second most common cancer in both women and men (after lung cancer) and the third most common if divided by gender (after breast and lung cancer for women and prostate and lung cancer for men). In the US alone, it accounts for 9% of all new cancer cases and causes 9% of all cancer deaths. This corresponds to 142,570 estimated new cases and 51,3700 deaths annually for US citizens (for 2010).[4] While the overall trend in colorectal cancer incidence is decreasing in the USA, rates are increasing among men and women under age 50 years.[5] In Europe, the situation is slightly different. With around 13,5% of all new cases and around 11,5-13,5% of all deaths (men and women), it is the second most common cancer for both genders combined and women alone (after breast cancer) and third for men (after lung and prostate) as well as causing the second most cancer deaths (after lung for men and breast for women). This means 435,600 expected new cases and 212,100 deaths per year for citizens of 40 European countries (for 2008) and 333,400 cases and 148,800 deaths annually for the European Union (27 countries) alone.[6] In Germany, 37,254 new cases and 13,748 deaths were observed per year (data from 2004, published 2010), which corresponds to second highest incidence for both genders (alone and combined, after prostate and breast cancer respectivly) and second highest mortality (both genders, but after lung cancer for men and breast cancer for women).[7] The differences between the USA and Europe in incidence and mortality can be partly explained by differences in lung cancer incidence (particularly lower for European women) probably caused by smoking habits.

1.2 Biology

Colorectal cancer is a disease originating from the epithelial cells lining the colon or rectum of the gastrointestinal tract. The colon is the last part of the digestive system - it extracts water and potassium salt and some fat soluble vitamins from solid wastes before they are eliminated from the body, and is the site in which flora-aided (largely bacteria) fermentation of unabsorbed material occurs. It does not play a major role in absorption of foods and nutrients.

The colon consists of four sections: the ascending colon, the transverse colon, the descending colon, and the sigmoid colon. The colon, cecum, and rectum make up the large intestine. Most intestinal cancers are located in the large intestine (about 80-90% of them are colorectal adenocarcinomas); other malignant carcinomas like gastrointestinal stroma tumors and neuroendocrine tumors are relatively rare (about 2% of all intestinal cancers). Cancers form more frequently further along the large intestine as the contents become more solid, therefore most cancer are located on the left side of the colon (about 60% vs. only 25% in the cecum and ascending colon), about half of those (approx. 55%) are in the sigmoid colon and rectum.[8]

Tumors are classified according to histological criteria (e.g. the size, the cell type from which the cancer originates and the grade of differentiation, see staging); the most common colon cancer cell type is adenocarcinoma which accounts for 90-95% of all cases. Adenocarcinomas are epithelial tumors, originating from glandular epithelium of the colorectal mucosa. The great majority of these adenocarcinomas arise sporadically (95%), only 5% are hereditary (also called Lynch syndrome or hereditary nonpolyposis colorectal cancer - HNPCC).[9] Subtypes are mucinous adenocarcinomas (up to 5%, when tumor cells are discohesive and secrete mucus, these large pools of mucus are optically "empty" spaces in histological samples) and signet-ring cell carcinomas (up to 1%, if the mucus remains inside the tumor cell, it pushes the nucleus at the periphery). Depending on glandular architecture, cellular pleomorphism, and mucosecretion of the predominant pattern, adenocarcinoma may present three degrees of differentiation: well, moderately, and poorly differentiated – the histological grade, which a pathology report will usually contain together with a description of cell type (see staging below). Other, rarer types of cancers in the large intestine include lymphomas and squamous cell carcinoma.[10] Many colorectal cancers are thought to arise from adenomatous polyps in the colon. These mushroom-like abnormal growths are usually benign, relatively well differentiated and non-invasive but some may become malignant and develop into adenocarcinoma over time (see risk factor section below for details). Familial adenomatous polyposis (FAP) is an inherited condition which leads to the development of hundreds to thousands of polyps in the colon due to mutations in the APC or MUTYH genes (see molecular changes below) and as a consequence of this mutation carriers have a near 100 percent lifetime risk of developing colorectal cancer (but this accounts for less than one percent of all cases).[11] This progress is thought to develop with increasing genetic instability and involving mutations which cause the evolution of invasive and metastasic tumours. This involves a series of alterations in oncogenes and tumour-suppressor genes that roughly follow the progression from small benign adenomas to advanced metastatic tumours. The stages of disease, from early to late, are directly related to the acquisition of these sequential genetic changes - the so called adenoma-carcinoma sequence, originally devised by Bert Vogelstein in the 1980s (see figure 2). [12, 13]

While polyps are attached to the mucous membrane (mucosa) in the colon, they do not penetrate it, in contrast to the full blown adenocarcinoma. When the tumor grows further, it invades the intestinal wall and thin layer of smooth muscle; infiltrating the submucosa (thereby reaching the lymphatic and blood vessels), and finally the muscular coat (muscularis propria). If it then acquires the ability to penetrate the walls of lymphatic and/or blood vessels it will metastasize, first to regional lymph nodes and then to other organs (mostly the liver, sometimes lung, rarely bone or brain). See also figure 3 in the staging section below for details. Localized colon cancer is usually diagnosed through colonoscopy (see at diagnosis below).[14]

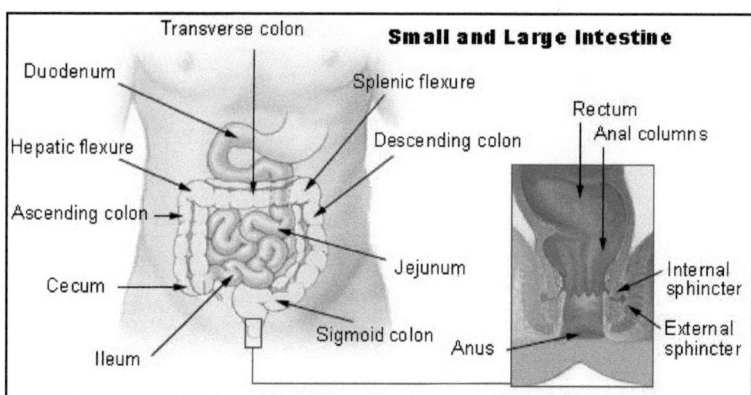

Figure 1 Intestinal anatomy (picture courtesy National Cancer Institute) - http://training.seer.cancer.gov/anatomy/digestive/regions/intestine.htm

1.2.1 Molecular changes

Colorectal cancers acquire many genetic changes, but certain signaling pathways are clearly singled out as key factors in tumor formation with activation of the Wnt signaling pathway as the key initiating event. The most common mutation in colorectal cancer inactivates the gene that encodes the adenomatosis polyposis coli (APC) protein. In the absence of functional APC — the brake on CTNNB1 (beta-catenin) — Wnt signaling is inappropriately and constitutively activated. Without APC, beta-catenin moves into the nucleus, binds to DNA, and activates more proteins. In a small subgroup of tumors with wild-type APC, mutations of beta-catenin that render the protein resistant to the beta-catenin degradation complex activate Wnt signaling. Somatic mutations and deletions that inactivate both copies of APC are present in most sporadic colorectal adenomas and cancers.[15] For CRC a large numbers of oncogenes and tumour suppressor genes including the Kirsten rat sarcoma-2 viral homologue (KRAS) and tumour protein 53 (TP53) are known.[15] Most colorectal cancer tumors are thought to be cyclooxygenase-2 (COX-2) positive. This enzyme is generally not found in healthy colon tissue, but is thought to fuel abnormal cell growth. However, it has been suggested that COX-2 protein expression is reduced in colorectal cancer with a defective mismatch repair (MMR) system (such as MUTYH and MLH1 mutations) which may explain the lack of response to COX-2 inhibiors.[16]

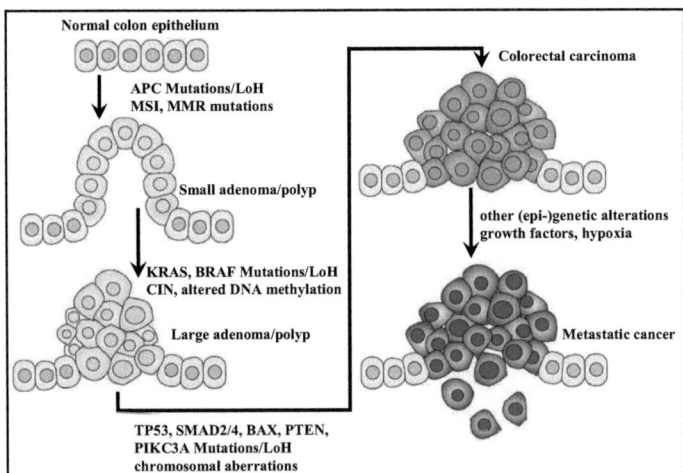

Figure 2 The colorectal adenoma–carcinoma sequence and alterations of genes that drive the progression of colorectal cancer (adapted from[13, 15] see also[12, 17]).

1.2.2 Risk Factors

While the life time risk for developing colorectal cancer is low (about 6% in Germany and 7% in the US), the risk increases dramatically with age. Most cases occur in the 60s and 70s, while cases before age 50 are uncommon unless a family history of early colon cancer is present. The mean age at diagnosis is about 65 years and older in Germany and the USA[18]. Family and personal history is a factor[19], as individuals who have previously been diagnosed and treated for cancer are at higher risk for developing colorectal cancer in the future. For example, women who have had cancer of the ovary, uterus, or breast are such a group.[20] As in most cancers, smokers are more likely to die of colorectal cancer than non-smokers, studies found a 30%-40% increase in risk of dying from the disease.[21] Alcohol consumption also increases the risk, possibly due to epigenetic mechanisms, as alcohol depletes the body of the methyl donor S-Adenosylmethionin (SAM), it may be a cause of earlier onset of colorectal cancer.[22] Polyps of the colon, particularly adenomatous polyps, are another risk factor for colon cancer. The removal of colon polyps at the time of colonoscopy reduces the subsequent risk of colon cancer.[23] As mentioned above, polyps are thought of as precursor lesions from which colorectal cancer arises, however some polyps are considered benign, depending on their type and size. The most common general classification is: hyperplastic polyps or serrated polyps, which are seen as mostly benign in nature (however, some histological subtypes are believed to be precancerous lesions); neoplastic polyps, which can be adenomatous (pre-malignant) or malignant and can be further broken down into subtypes depending on the histological growth pattern; hyperplastic polyps which carry little malignant potential and inflammatory polyps, associated with inflammatory conditions such as Ulcerative Colitis and Crohns disease.[24] Adenomas (adenomatous polyps) carry the greatest malignant potential, depending on their subtype and/or size: 5% risk of cancer if the adenoma is tubular, 20% risk of cancer if it is tubulovillous and 40% risk of cancer if villous; if the polyp is greater than 1 cm the cancer risk is about 10% while if it is greater than 2 cm the risk rises to 15%. Therefore, normally a polyp which is greater than 0.5 cm is removed during a colonoscopy regardless of its type.[25]

Inflammatory bowel diseases (such as ulcerative colitis and Crohns disease which have similar symptoms but differ in pathology and involvement of intestinal parts) can lead to an increased risk, but only one percent of colorectal cancer patients have a history of chronic inflammation (in case of colitis, for Crohns disease this is even lower).[26]

In recent years, a lot of attention has been paid to environmental factors, since people living in industrialized countries are at a relatively increased risk compared to less developed countries with high-fiber/low-fat diets. Overall, studies suggests that a diet high in red meat and low in fresh fruit, vegetables, poultry and fish as well as low in fiber, are associated with an increased risk of colorectal cancer. Other studies dispute that diets high in fiber decrease the risk of colorectal cancer, rather, low-fiber diet was associated with other risk factors, leading to confounding, so the nature of the relationship between dietary fiber and risk of colorectal cancer remains controversial.[27-29] Vitamin B6 intake is inversely associated with the risk of colorectal cancer.[30] Body weight, physical activity and fat absorption also play a role in disease development, linking obesity to colorectal cancer.[31] Exposure to some viruses (such as particular strains of human papilloma virus) may be associated with colorectal cancer.[32]

1.2.3 Diagnosis

Colorectal cancer can take many years to develop and early detection of colorectal cancer greatly improves the chances of a cure. Despite this, colorectal cancer screening rates remain low, though screening for the disease is recommended in individuals who are at increased risk or over 50 years of age. The most used methods are Fecal occult blood test (FOBT)[33], a test for blood in the stool and colonoscopy[34]. Two main types of tests can be used for detecting occult blood in stools i.e. guaiac based (gFOBT) and immunochemical (FIT). The gFOBT works by detecting the heme component in hemoglobin through its peroxidase-like activity, in effect rapidly breaking down hydrogen peroxide. In contrast, the FIT uses specific antibodies to detect globin. There are various commercially available test kits which have been categorized as being of low or high sensitivity, and only high sensitivity tests are now recommended in colon cancer screening. In general, the sensitivity of immunochemical testing is superior to that of chemical testing without an unacceptable reduction in specificity. [33] The sensitivity of a single stool guaiac test to pick up bleeding has been quoted at 10 to 30%, but if a standard three tests are done as recommended the sensitivity rises to 92%.

Another method, fecal porphyrin quantification (also called Hemoquant) permits precise quantification of hemoglobin after conversion to porphyrin by comparative fluorescence against a reference standard. Several pitfalls exist, as e.g. optimal clinical performance of the stool guaiac test depends on preparatory dietary adjustment and for immunochemical tests the number of fecal samples submitted may affect the clinical sensitivity and specificity of the methodology. [35] Also, the detection rate of the test decreases if the time from sample collection to laboratory processing is delayed. [36]

For colonoscopy, a lighted probe called a colonoscope is inserted into the rectum and the entire colon to look for polyps and other abnormalities that may be caused by cancer. A colonoscopy has the advantage that if polyps are found during the procedure they can be removed immediately. Tissue can also be taken for biopsy. Most colorectal cancers should be preventable, theoretically, through increased surveillance and removal of polyps. It has been suggested, that the risk of cancer death would decrease by > 80%, provided that colonoscopy is started by the age of 50, and repeated every 5 or 10 years. In the United States, colonoscopy or FOBT plus sigmoidoscopy are the preferred screening options, but only every 5 to 10 years, in average risk individuals. In Germany, periodic colonoscopy starting by the age of 55 is paid by the health insurance companies since 2002 as a prophylactic measure for colorectal cancer prevention. FOBT tests are paid beginning with the age of 50 years. If the testing is negative (i.e. no polyps are found), this can be repeated in 10 year intervals (but if the patient develops symptoms in between, another colonoscopy is possible and paid for). The data from 2002 to 2005 shows that in about 20% of all colonoscopies a polyp was found, a tumor in less then 1% (and those tumors were mostly early stages I and II, see staging below). However, only about 10% of the eligible persons took advantage of the procedure, thus overall compliance is low, possibly due to the invasiveness of the procedure.[37] According to a randomized trial in the United Kingdom, a one-time flexible colonoscopy screening between 55 and 64 years of age reduced colorectal cancer incidence by 33% and mortality by 43%.[38] Stool DNA testing is an emerging technology in screening for colorectal cancer. Premalignant adenomas and cancers shed DNA markers from their cells which are not degraded during the digestive process and remain stable in the stool. Capture, followed by PCR amplifies the DNA to detectable levels for assay. Clinical studies have shown a cancer detection sensitivity of 71%–91%.[39] Most of the studies of stool-based DNA biomarkers have focused on the detection of aberrant DNA originating from colorectal cancers using (epi-)genetic alterations such as mutations and hypermethylation of the promoter region of specific genes (see also next chapter of this thesis). A number of markers have been studied, such as mutations in KRAS and APC genes, methylation of SFRP2, TFPI2, GATA4, NDRG4, OSMR, and VIM—with no marker emerging as obviously the best. The value of a panel of multiple markers has also been evaluated using combinations of several gene variants with and without markers of MSI (microsatellite instability), methylation, and/or DNA integrity. As a generalization, sensitivities for cancer ranged from around 40% to almost 90% and specificity tended to be in the range of 90% to 95%. Such DNA tests have great potential as they tend to be more sensitive than fecal blood testing, including detection of early stage disease, when treatment is most effective but as yet no clearly highly effective panel currently exists. [34, 40]

Blood tests for protein markers such as carcinoembryonic antigen (CEA) are frequently false positive or false negative, and are not recommended for screening, but can be useful to assess disease recurrence.[41] In contrast, DNA shed from tumors circulating as cell free DNA in the blood can be detected in the plasma of colon cancer patients using highly sensitive assays for either DNA mutations or DNA methylation tumor markers (see also the next chapter of this thesis). [42, 43]

1.2.4 Staging

The most common staging system for colorectal cancer[44] (and most other tumors as well) in use is the TNM (for tumors/nodes/metastases)[45] system, developed and maintained by the Union for International Cancer Control (UICC) in Europe and also used by the American Joint Committee on Cancer (AJCC)[46] in the USA (see **Table 1**). As all cancer staging systems it is a description (with roman numerals I to IV) of the extent the cancer has spread - i.e. the progression of the disease. Originally developed in France (by Pierre Denoix between 1943 and 1952), it is now used as a globally recognised standard (in 1987, the UICC and AJCC staging systems were unified into a single staging system) in most hospitals and cancer registries. The TNM system assigns a number based on three categories. "T" denotes the degree of invasion of the intestinal wall (on a scale from 1-4), "N" the degree of lymphatic node involvement (range 0-3), and "M" the degree of metastasis (0 or 1). The broader stage of a cancer is usually quoted as a number I, II, III, IV derived from the TNM value grouped by prognosis; a higher number indicates a more advanced cancer and likely a worse outcome. While the TNM parameters are mandatory, the histological grade of the tumor cells is usually assessed as well, classified as "G" (ranging from 1-4). The grade score increases with the lack of cellular differentiation - it reflects how much the tumor cells differ from the cells of the normal tissue they have originated from, i.e. they are "low grade" if they appear similar to normal cells, and "high grade" if they appear poorly differentiated.

Additional Staging parameters can include: "R" (0-2) the completeness of the resection, if the patient was operated and the tumor removed; "V" (0-2) invasion into veins and "L" (0-1) invasion into lymphatic vessels. The TNM status can also be modified with a prefix, denoting if the stage was given by clinical examination (c) or by pathologic examination of a surgical specimen (p) as well as if it was assessed before or after therapy (y).

Introduction - Colorectal Cancer

Table 1 Colorectal cancer classification systems

AJCC/UICC	TNM stage	Dukes	MAC	TNM stage criteria for CRC
Stage 0	Tis N0 M0	-	-	Tis: Tumor confined to mucosa; cancer-in-situ
Stage I	T1 N0 M0	A	A	T1: Tumor invades submucosa
Stage I	T2 N0 M0	A	B1	T2: Tumor invades muscularis propria
Stage II-A	T3 N0 M0	B	B2	T3: Tumor invades subserosa or beyond (without other organs involved)
Stage II-B	T4 N0 M0	B	B3	T4: Tumor invades adjacent organs or perforates the visceral peritoneum
Stage III-A	T1-2 N1 M0	C	C1	N1: Metastasis to 1 to 3 regional lymph nodes. T1 or T2.
Stage III-B	T3-4 N1 M0	C	C2/C3	N1: Metastasis to 1 to 3 regional lymph nodes. T3 or T4.
Stage III-C	any T, N2 M0	C	C1/C2/C3	N2: Metastasis to 4 or more regional lymph nodes. Any T.
Stage IV	any T, any N, M1	D	D	M1: Distant metastases present. Any T, any N.

An older and less complicated staging system that predates the TNM system is the Dukes Classification, developed in 1932 by the British pathologist Cuthbert Dukes. It has only 4 stages (A-D) depending on the invasiveness of the tumor (ranging from invasion into the membrane with or without involvement of lymph nodes to widespread metastasis)[47]. An adaptation by the Americans Astler and Coller in 1954 further divided stages B and C into substages (B1-2 and C1-2); this is the modified Astler-Coller classification (MAC). Although several different forms of the Dukes classification were developed, however this system has largely be replaced by TNM (see figure below).[48]

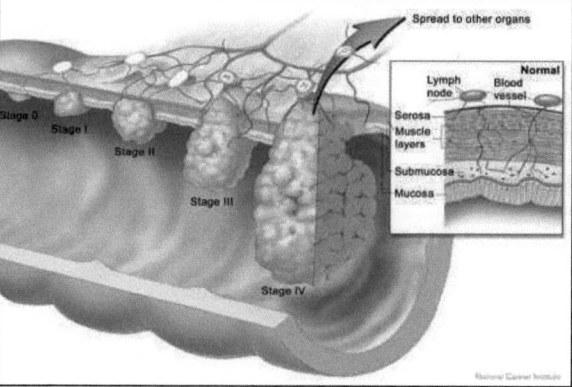

Figure 3 Graphical overview of colorectal cancer staging (picture courtesy National Cancer Institute) - www.ccalliance.org/what_diseaseinfo_staging.html

1.2.5 Treatment

Treatment depends on cancer stage, colorectal cancer can be curable at early stages, when it is detected at later stages it is less likely to be curable. If the tumor is localized, curative surgical treatment can be possible, but a more advanced tumor typically requires surgical removal of the section of colon containing the tumor with sufficient margins, i.e. radical resection of surrounding tissue and lymph nodes to reduce local recurrence.

Surgery remains the primary treatment while chemotherapy and/or radiotherapy may be recommended depending on the individual patient's staging and other medical factors. Because colon cancer primarily affects the elderly, it can be a challenge to determine how aggressively to treat a particular patient, especially after surgery. Clinical trials suggest that "otherwise fit" elderly patients fare well if they have adjuvant chemotherapy after surgery, so chronological age alone should not be a contraindication to aggressive management.[49] Surgeries can be categorised into the categories curative, palliative or bypass. Curative surgical treatment can be offered if the tumor is localized, like very early cancer that develops within a polyp (by removing the polyp during colonoscopy - polypectomy)[50] or stages I-II, this typically requires surgical removal of the section of colon containing the tumor with sufficient margins, and radical en-bloc resection of mesentery and lymph nodes to reduce local recurrence (colectomy). The remaining parts of colon are the reconnected (anastomosis) to preserve a functioning colon, if this is not possible a stoma (artificial orifice) is created.[51]

Noncurative (palliative) resection of the primary tumor is still offered to reduce further morbidity caused by tumor bleeding, invasion, and its catabolic effect. Surgical removal of isolated liver metastases is common and may be curative in selected patients; improved chemotherapy has increased the number of patients who are offered surgical removal of isolated liver metastases.[52] If the tumor invaded into adjacent vital structures, which makes excision technically difficult, the surgeons may prefer to bypass the tumor (ileotransverse bypass) or to do a proximal fecal diversion through a stoma.[53]

Chemotherapy is often applied after surgery (adjuvant), before surgery (neoadjuvant), or as the primary therapy (palliative). For colorectal cancer, chemotherapy after surgery is usually only given if in advanced stages if the cancer has spread to the lymph nodes (Stage III-IV). Chemotherapy is used to reduce the likelihood of metastasis developing, shrink tumor size, or slow tumor growth.[54] Adjuvant (after surgery) chemotherapy usually consists of (5-)fluorouracil (5-FU, a pyrimidine analog and antimetabolite which incorporates into the DNA molecule and stops synthesis, sold under the brand names Adrucil, Carac, Efudix, Efudex and Fluoroplex) or Capecitabine (trade name

Xeloda) which is the oral prodrug of 5-FU together with Leucovorin (LV, Folinic Acid, a vitamin B derivative that modulates/reduces the side effects of fluorouracil) and Oxaliplatin (trade name Eloxatin, an alkylating agent which inhibits DNA synthesis). This is known as the FOLFOX regimen.[55] In a neoadjuvant setting (i.e. chemotherapy for metastatic disease) commonly used first line chemotherapy regimens involve the combination of infusional 5-fluorouracil, leucovorin, and oxaliplatin (i.e. FOLFOX) with bevacizumab (trade name Avastin, a humanized monoclonal antibody that recognizes and blocks vascular endothelial growth factor A, VEGF-A) or FOLFIRI, which is infusional 5-fluorouracil, leucovorin, and irinotecan (a topoisomerase inhibitor, which prevents DNA from uncoiling and duplicating) with bevacizumab or the same chemotherapy drug combinations with cetuximab (a chimeric, i.e. mouse/human monoclonal antibody, inhibits epidermal growth factor receptor, EGFR) in KRAS wild type tumors.[56] Second line regimens usually switch then from either FOLFOX to FOLFIRI or vice versa. Colorectal cancer patients that have a mutation in the KRAS gene do not respond to Erbitux (cetuximab) and Vectibix (panitumumab, a fully human monoclonal antibody specific to the EGFR), therefore patients are now tested for KRAS gene mutations before being offered these EGFR-inhibiting drugs.[57] Radiotherapy is not used routinely in colon cancer, as it could lead to radiation enteritis, and it is difficult to target specific portions of the colon. It is more common for radiation to be used in rectal cancer, since the rectum does not move as much as the colon and is thus easier to target. However, as a palliative measure it might be used in colon cancer for pain relieve targeted at metastasis if they compress vital structures and/or cause pain. Chemotherapy agents are used to increase the effectiveness of radiation by sensitizing tumor cells if present. Radiation therapy is used in rectal cancer in an neoadjuvant setting[58] (together with chemotherapy - chemoradiation), given before surgery in patients with tumors that extend outside the rectum or have spread to regional lymph nodes, in order to decrease the risk of recurrence following surgery or to allow for less invasive surgical approaches.[59] In an adjuvant setting - where a tumor perforates the rectum or involves regional lymph nodes (stage III-IV) and for palliative care to decrease the tumor burden in order to relieve or prevent symptoms.[60] Generally speaking, invasive cancers that are confined within the wall of the colon (tumor–node–metastasis stages I and II) are curable, but if untreated, they spread to regional lymph nodes (stage III) and then metastasize to distant sites (stage IV). Stage I and II tumors are usually curable by surgical excision, and up to 73% of cases of stage III disease are curable by surgery combined with adjuvant chemotherapy. Recent advances in chemotherapy have improved survival, but stage IV disease is usually incurable.[61] Up to 25% of patients with metastatic (stage IV) colorectal cancer at the time of diagnosis (mostly in liver and lymph nodes,

sometimes also lung metastases) will have isolated liver metastasis that is potentially resectable. Lesions which undergo curative resection have demonstrated 5-year survival outcomes now exceeding 50%.[62] Lesions confined to the right lobe are amenable to en bloc removal with a right hepatectomy (liver resection) surgery. Smaller lesions of the central or left liver lobe may sometimes be resected in anatomic "segments", while large lesions of left hepatic lobe are resected by a procedure called hepatic trisegmentectomy. Treatment of lesions by smaller, non-anatomic "wedge" resections is associated with higher recurrence rates. Some lesions which are not initially amenable to surgical resection may become candidates if they have significant responses to preoperative chemotherapy or immunotherapy regimens. Lesions which are not amenable to surgical resection for cure can be treated with modalities including radio-frequency ablation (RFA), cryoablation, and chemoembolization.[63] Patients with colon cancer and metastatic disease to the liver may be treated in either a single surgery or in staged surgeries (with the colon tumor traditionally removed first) depending upon the fitness of the patient for prolonged surgery, the difficulty expected with the procedure with either the colon or liver resection, and the comfort of the surgery performing potentially complex hepatic surgery.[64]

Prognosis

As mentioned above, survival depends on the stage - survival rates for early stage detection are about 5 times that of late stage cancers. If the cancer is symptomatic, it is typically quite advanced and overall survival is poor, but it depends on the type and if it has already spread to other organs. For example, a tumor that hasn't breached the muscular layer - stage I has an average five-year-survival of approximately 85%-90%. This drops to about 50-60% for a more invasive tumor (stage II), yet without node involvement it can reach 70%. Cancers with positive regional lymph nodes (stage III) have an average 5-year survival of 30%-40%, while for cancers with distant metastases (stage IV) is only about 5%.[61]

Table 2 Survival Rates: Colorectal Cancer (according to the National Cancer Institute)

Stage	Survival Rate
Stage 0	>96%
Stage I	80-95%
Stage II	55-80%
Stage III	35-55% involved lymph nodes
Stage IV	< 15% distant metastases

Follow-up and monitoring

The aim of follow-up is the diagnosis of any metastasis or tumors that develop later but did not originate from the original cancer (metachronous lesions), i.e. recurrence of the disease as early as possible.[65] Guidelines recommend examinations every 3 to 6 months for 2 years, then every 6 months for 5 years.[66]

Carcinoembryonic antigen (CEA) is a protein found on virtually all colorectal tumors. CEA may be used to monitor and assess response to treatment in patients with metastatic disease. CEA can also be used to monitor recurrence in patients post-operatively. CEA blood level measurements are only advised for patients with T2 or greater lesions who are candidates for intervention. Imaging methods such as CT (X-ray computed tomography) scans or Magnetic resonance imaging (MRI), Positron emission tomography (PET) are considered annually for the first 3 years for patients who are at high risk of recurrence and are candidates for curative surgery if recurrence indeed occurs. Colonoscopy can be done after 1 year, except if it could not be done during the initial staging because of an obstructing mass, in which case it should be performed after 3 to 6 months. If a villous polyp, polyp >1 centimeter is found, it can be repeated after 3 years, then every 5 years. These guidelines are based on recent meta-analyses showing that intensive surveillance and close follow-up can reduce the 5-year mortality rate from 37% to 30%.[67-69]

1.3 Epigenetics and Epigenomics

Epigenetics is defined as heritable changes in phenotype or gene expression that are not accompanied by or occur independent of changes in the DNA sequence but caused by other mechanisms hence the name *epi-* (the Greek word meaning over, above, on top of). These changes may remain through cell divisions for the remainder of the cell's life and may also last for multiple generations since they are preserved when cells divide. The word was coined by C. H. Waddington in 1942 as a portmanteau of the words genetics and epigenesis and in 1990 defined by Robin Holliday as "the study of the mechanisms of temporal and spatial control of gene activity during the development of complex organisms." Thus epigenetic can be used to describe anything other than DNA sequence that influences the development of an organism. While the definition initially referred only to the role of epigenetics in embryonic development; however, the definition of epigenetics has evolved over time as it is implicated in a wide variety of biological processes.[70-72]

The molecular basis of epigenetics involves modifications of the activation of certain genes through chemical modifications of individual amino acids on the tails of proteins called histones. Additionally, the chromatin proteins associated with DNA may be activated or silenced. This accounts for why the differentiated cells in a multi-cellular organism express only the genes that are necessary for their own activity. Most of these heritable changes are established during differentiation and are stably maintained through multiple cycles of cell division, enabling cells to have distinct identities while containing the same genetic information. The heritability of gene expression patterns is mediated by epigenetic modifications, which include methylation of cytosine bases in DNA, posttranslational modifications of histone proteins as well as the positioning of nucleosomes along the DNA.[73-75] Gene silencing at the level of chromatin is necessary for the life of eukaryotic organisms and is particularly important in orchestrating key biological processes, including differentiation, imprinting, and silencing of large chromosomal domains. To date, the best studied epigenetic transcriptional control mechanisms involve DNA methylation and covalent modifications of histone proteins.[76-78]

The complement of these modifications – the overall epigenetic state of a cell, collectively referred to as the epigenome, provides a mechanism for cellular diversity by regulating what genetic information can be accessed by cellular machinery. In many species, this regulation can be initiated and maintained solely by processes involving the covalent modifications of histones and other chromatin components. In effect, this means that individuals have a single genome but many "epigenomes".[79]

The phrase "epigenetic code"[80] has been used to describe the set of epigenetic features that create different phenotypes in different cells, this could represent the total state of the cell, with the position of each molecule accounted for in an *epigenomic map*, a diagrammatic representation of the gene expression, DNA methylation and histone modification status of a particular genomic region.[81] More typically, the term is used in reference to systematic efforts to measure specific, relevant forms of epigenetic information such as the histone modifications or DNA methylation patterns.[82] The interactions between DNA methylation and histone modifying enzymes further enhance the complexity of epigenetic regulation of gene expression, which determines and maintains cellular identity and function (see figure below).[83]

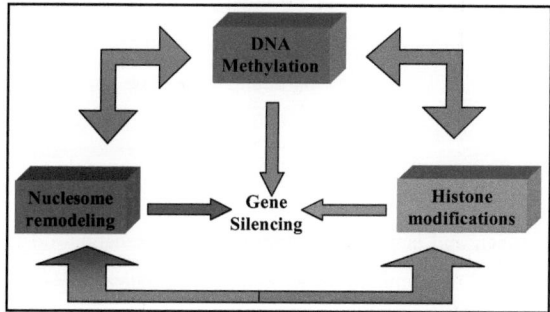

Figure 4 Interplay between DNA methylation, histone covalent modifications, and nucleosomal remodeling. Adapted from Jones.[71]

1.3.1 Histone modifications and the histone code

Chromatin is made of repeating units of nucleosomes, which consist of around 147 base pairs of DNA wrapped around an octamer of four core histone proteins (H3, H4, H2A and H2B) and a linker histone, (H1, which binds the nucleosome and the entry and exit sites of the DNA and interacts with the linker DNA between nucleosomes - thus locking the DNA into place). This organisation helps in the compaction of chromatin into higher order structures. In addition to serving as the basic modules for DNA packaging within a cell, nucleosomes regulate gene expression by altering the accessibility of regulatory DNA sequences to transcription factors.[75] Histone proteins are a group of closely related proteins that vary often by only a few amino acids encoded in multiple intronless genes mostly found in the large histone gene cluster on chromosome 6 in humans. The transcripts lack polyA tails but instead contain a palindromic termination element.

For example, for histone H4 there are 12 gene variants known so far (HIST1H4A-L and HIST4H4). While histones are highly conserved, several variant forms have been identified. Interestingly, this diversification of histone function is restricted to H2A and H3, with H2B and H4 being mostly invariant. Notable variants include H2A.1, H2A.2, H2A.X, H2A.Z and H3A1, H3A2, H3A3 (for H4 only 2 variants are known, H4.1 and H4.2). The incorporation of histone variants into nucleosomes also influences nucleosome occupancy and thus gene activity, e.g. H3A3 and H2A.Z are preferentially enriched at promoters of active genes or genes poised for activation and can mediate gene activation by altering the stability of nucleosomes.[73, 84]

As mentioned before, post-translational modifications of histones play important roles in controlling and maintaining chromatin structure and form a major category of epigenetic transcriptional control. The histone tails on the nucleosome surface are subject to such post-translational modifications that may form a code specifying patterns of gene expression by helping to determine the transcriptional activity of a particular gene. The complement of modifications is proposed to store the epigenetic memory inside a cell in the form of a 'histone code' that determines the structure and activity of different chromatin regions.[72, 85]

These covalent modifications of histone proteins can change densely compacted, inactive heterochromatin to the open and active configuration of euchromatin, and vice versa. Subject to covalent modification, including acetylation and methylation, which may alter expression of genes located on DNA associated with its parent histone octamer, are N-terminal histone tails on H4 and H3. These modifications include the covalent attachment of methyl or acetyl groups to lysine and arginine amino acids and the phosphorylation of serine or threonine.[73, 86]

Each of these modifications can be subjected to further variations that can change its function. For instance, methylation of arginine can involve the addition of 1, 2 or 3 methyl groups, each conferring subtly different functional consequences (as Di- and Tri-methylation of Lysine 9 on H3 are associated with repression and heterochromatin, while mono-methylation of K9 is associated with active genes). Histone modifications work by either changing the accessibility of chromatin or by recruiting and occluding non-histone effector proteins, which decode the message encoded by the modification patterns. For example, two methylation events involving lysines 9 and 27 on the N-terminus of histone H3 (H3-K9m3 and H3-K27m3) are associated with transcriptional silencing and repression, whereas trimethylation of lysine 4 of histone H3 (H3-K4m3) is associated with an open chromatin structure surrounding promoter regions of active genes.[75, 87]

The mechanism of inheritance of the histone code is still not fully understood. The histone modifications, which mark both active and inactive chromatin, are made possible by several families of enzymes, besides histone methyltransferases (HMTs) and histone demethylases (HDMs); this includes both histone acetyltransferases (HATs) and histone deacetylases (HDACs). These histone-modifying enzymes interact with each other as well as other DNA regulatory mechanisms to tightly link chromatin state and transcription. Generally, the active chromatin structure corresponding to increased transcriptional activity is associated with increased histone acetylation, logically, acting antagonistically to HATs, HDACs produce transcriptional repression. In this regard, acetylation of key histone amino acid residues has been most extensively studied over the past few years and is maintained by a balance between the activities of histone acetyltransferases and histone deacetylases. Since HDACs work through a complicated mechanism that involves interaction with DNA methyltransferases and methyl-CpG-binding proteins and transcriptionally repressing histone marks such as H3K9me3 and H3K27me3 work in concert with DNA methylation the two main silencing mechanisms in mammalian cells are thus depending on each other. This is also true for histone methylation, since several HMTs, including can direct DNA methylation to specific genomic targets by directly recruiting DNA methyltransferases (DNMTs) to stably silence genes and in addition to the direct recruitment, HMTs and demethylases also influence DNA methylation levels by regulating the stability of DNMT proteins (see figure below).[73, 75, 85, 88]

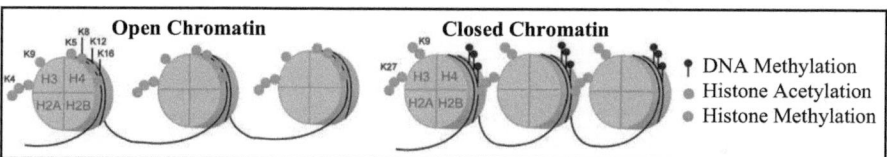

Figure 5 Histone modifications and their influence on chromatin formation (as well as their interplay with DNA methylation marks).[89]

1.3.2 DNA Methylation

DNA methylation is found in the genomes of diverse organisms including both prokaryotes and eukaryotes. It involves the addition of a methyl group to the 5 position of the cytosine pyrimidine ring or the number 6 nitrogen of the adenine purine ring. This addition is catalyzed through an enzymatic reaction which uses S-adenosyl-methionine (SAM) as a methyl group donor and action of DNA methyltransferases (see figure below). In prokaryotes, DNA methylation occurs on both cytosine and adenine bases and encompasses part of the host restriction system. In eukaryotes, however, methylation seems to be confined to cytosine bases and is associated with a repressed chromatin state and inhibition of gene expression. Especially vertebrates, have taken advantage of the heritability of DNA cytosine methylation patterns to add another layer of control to epigenetic silencing processes. DNA methylation at the 5 position of cytosine has been found in every vertebrate examined.[75, 90, 91]

In the mammalian genome, methylation of cytosine residues occurs most commonly at CpG (the "p" in CpG refers to the phosphodiester bond between the cytosine and the guanine, which indicates that the C and the G are next to each other on the sequence strand and not triple-bonded across two strands of DNA) dinucleotides and occasionally at CA or CT residues (non-CpG methylation is prevalent in embryonic stem cells[92]). In adult somatic tissues, DNA methylation typically occurs in a CpG dinucleotide context; non-CpG methylation is prevalent in embryonic stem cells. The resultant base, 5-methylcytosine, is relatively unstable, and prone to spontaneous deamination to form thymine. Therefore, the distribution of CpG dinucleotides in the human genome is not uniform, probably due to the high mutagenic potential of 5-methylcytosine. While there exits the Thymine-DNA glycosylase enzyme in humans (TDG) that specifically replaces T's from T/G mismatches, it is not sufficiently effective to prevent the relatively rapid mutation of the dinucleotides. If the spontaneous deamination of 5-methylcytosine to form the DNA base thymidine is not recognized and repaired, a cytosine-to-thymidine change remains, in this way, DNA methylation can be seen as an endogenous mutagen. This process has resulted in a progressive depletion of CpG dinucleotides over the course of time (in humans, only one to two percent of the human genome are CpG clusters, which means CpGs occur at only 21% of the expected frequency), a phenomenon called CG suppression. Between 60% and 90% of all CpGs are methylated in mammals. This methylation in normal cells may contribute to maintaining the large amount of noncoding DNA in a transcriptionally inert state, since most of the CpG dinucleotides, which are not associated with promoter regions, are predominantly methylated. As a result, CpGs are relatively rare unless there is selective pressure to keep them or a region is not methylated for some reason, like having to do with the regulation of gene expression.[73, 86, 70, 76, 93] In humans, about 70-90% of all CpG dinucleotides in the human genome are heavily methylated (depending on the tissue and cell type)[94] and the remainder are typically seen in CpG-rich regions that span the promoters and sometimes the first exons of genes and are normally unmethylated. These regions, known as CpG islands, are found in association with about 60% of all human genes, mostly covering promoter regions (older estimates were about 40%), are typically 300-3,000 base pairs in length and usually defined with an observed CpG/expected CpG of 0.65 – this means they are made up of about 65% CG residues.[73, 91, 95, 96] Based on an extensive search on the complete sequences of human chromosomes 21 and 22, DNA regions >500 bp with a GC content >55% and observed CpG/expected CpG of 0.65 were more likely to be the true CpG islands associated with the 5' regions of genes.[97] However, DNA methylation does not occur exclusively at CpG islands. Recently, the term "CpG island shores" was coined for regions of lower CpG density that lie in close proximity (~2kb) of CpG islands.[98] The methylation of these CpG island shores is closely associated with transcriptional inactivation. Most of the tissue-specific DNA methylation seems to occur not at CpG islands but at CpG island shores. Differentially methylated CpG island shores are sufficient to distinguish between specific tissues and are conserved between human and mouse. Moreover, 70% of the differentially methylated regions in reprogramming are associated with CpG island shores.[91, 99]

The lack of methylation in promoter-associated CpG islands permits expression of a gene, if the appropriate transcription factors are present, and the chromatin structure allows access to them. The majority of CpG islands usually remain unmethylated during development and in differentiated tissues, particularly housekeeping genes. The exception to this unmethylated state of CpG islands involves the silenced gene alleles for imprinted genes and genes encompassed within regions of X-chromosome inactivation, and this indicates the tight association of promoter DNA methylation and transcriptional silencing during normal mammalian development.[71, 73, 95, 96, 100]

Some tissue-specific CpG island methylation has also been reported to occur in a variety of somatic tissues, primarily at developmentally genes. The orchestration of methylation in CpG islands by an assortment of methylating and demethylating enzymes is thought to provide one of the layers of epigenetic control of germ-line and tissue-specific gene expression.[75, 101]

The repetitive genomic sequences that are scattered all over the human genome (like ALU and LINE elements) are heavily methylated, which prevents chromosomal instability by silencing non-coding DNA and transposable DNA elements and also endoparasitic and retroviral transposons.[102-104] The observation of the inverse relationship between CpG methylation and transcriptional activity and that this CpG methylation is associated with a repressed chromatin state originally led to the speculation that methylation of CpG sites in the promoter of a gene may inhibit gene expression. While the role of CpG island promoter methylation in gene silencing is well established, much less is known about the role of methylation of non-CpG island promoters. Some studies have shown that DNA methylation is also important for the regulation of non-CpG island promoters.[96, 105]

There are two general mechanisms by which DNA methylation inhibits gene expression: first, modification of cytosine bases can inhibit the association of some DNA binding factors with their cognate DNA recognition sequences by physically impeding binding and second, proteins that recognize methyl-CpG can elicit the repressive potential of methylated DNA. Methyl-CpG-binding domain proteins (MBDs) bind to methylated DNA and then recruit additional proteins to the locus, such as such as histone deacetylases and other chromatin remodeling proteins, which function as transcriptional co-repressor molecules to silence transcription and to modify surrounding chromatin to its silent state, therefore providing a link between DNA methylation and chromatin remodelling and modification.[106, 107]

This shows a link between DNA methylation and chromatin structure. Furthermore, DNA methyltransferases co-localize with heterochromatin and interact with methyl-CpG binding proteins as well as recruit HDACs to achieve gene silencing and chromatin condensation.[73, 75]

DNA methylation is also associated with histone modifications such as the absence of histone H3K4 methylation and the presence of H3K9 methylation, thereby establishing a repressive chromatin state.[108, 109]

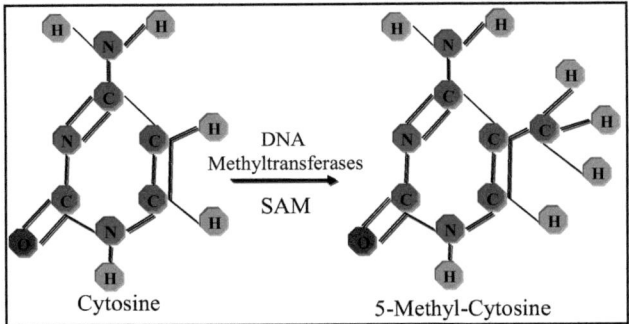

Figure 6 Methylation of cytosine in DNA occurs at CG dinucleotides. Methylation occurs by the addition of a methyl group at the 5' site of cytosine by DNA methyltransferases (DNMTs). Cytosine residues in DNA are thus converted to 5-methylcytosine The methyl group is donated by the universal methyl donor S-adenosylmethionine (SAM).[76, 77]

DNA Methyltransferases

In mammals, DNA methylation is regulated by a family of cytosine DNA methyltransferase enzymes (DNMTs) that includes DNMT1, DNMT3A and DNMT3B as well as DNMT3L. These proteins fit into two general classes based on their preferred DNA substrate: de novo (DNMT3a/b) and maintenance (DNMT1) methyltransferases, which in combination constitute the core enzymatic components of the DNA methylation system in mammals.[73, 75, 77, 110]

DNMT1is the most abundant DNA methyltransferase in mammalian cells, which acts during replication preferentially methylating hemimethylated DNA, is generally described as the maintenance DNA methyltransferase, which copies pre-existing methylation patterns onto the new DNA strand during DNA replication. Maintenance methylation activity is necessary to preserve DNA methylation after every cellular DNA replication cycle. Without it, the replication machinery itself would produce daughter strands that are unmethylated and, over time, would lead to passive demethylation.[111] Mouse models with both copies of DNMT1 deleted are embryonic lethal at approximately day 9-11, due to the requirement of DNMT1 activity for development in mammalian cells.[112] DNMT1 has several isoforms, the somatic DNMT1, a splice variant (DNMT1b) and an oocyte-specific isoform (DNMT1o). DNMT1o is synthesized and stored in the cytoplasm of the oocyte and translocated to the cell nucleus during early embryonic development, while the somatic DNMT1 is always found in the nucleus of somatic tissue.[113-115]

The precise DNA methylation patterns found in the genome are generated by the cooperative activity of the two de novo methyltransferases — DNMT3A and DNMT3B, which act independent of replication and show equal preference for both unmethylated and hemimethylated DNA. DNMT3A and 3B are important for patterning of DNA methylation during embryogenesis and early in development, since loss of each DNTM is embryonic lethal in knockout mice models. DNMT3a methylates CpG sites at a rate much slower than DNMT1, but greater than DNMT3b. [73, 75, 91]

DNMT3L is a protein that is homologous to the other two DNMT3s but has no intrinsic DNA methyltransferase catalytic activity. Instead, DNMT3L physically associates with DNMT3a and DNMT3b and assists them by increasing their ability to bind to DNA and modulating their catalytic activity. DNMT3L is also expressed during gametogenesis when genomic imprinting takes place. The loss of DNMT3L leads to bi-allelic expression of genes normally not expressed by the maternal allele.[116, 117]

However, the separable roles of the DNMT enzymes have been challenged in cancer models, some investigators have shown that severe depletion of DNMT1 produces (a) negligible decreases in overall DNA methylation and promoter methylation and (b) undetectable changes in expression of silenced tumor-suppressor genes. Therefore, in human cancer cells DNMT1 is responsible for both de novo and maintenance methylation of tumor suppressor genes. This fits to the observation that while it predominantly methylates hemimethylated CpG di-nucleotides in the mammalian genome and is 7–100 fold more active on hemimethylated DNA as compared with unmethylated substrate in

Introduction - Epigenetics

vitro, but it is still more active at de novo methylation than other DNMTs. DNMT3s can also interact with DNMT1, which might be a co-operative event during DNA methylation.[91]

To note, DNMT2 (TRDMT1) has been identified as a DNA methyltransferase homolog, containing the sequence motifs common to all DNA methyltransferases; however, DNMT2 (TRDMT1) does not methylate DNA but instead methylates cytosine-38 in the anticodon loop of aspartic acid transfer RNA. To reflect this different function, the name for this methyltransferase has been changed from DNMT2 to TRDMT1 (tRNA aspartic acid methyltransferase 1) to better reflect its biological function. TRDMT1 is the first RNA cytosine methyltransferase to be identified in a human.[118-120]

Finally, DNMT3a and DNMT3b are targets for several micro-RNAs (especially interesting in the context of colorectal cancer) but miRNAs in general, like normal genes, can be regulated by epigenetic mechanisms such as DNA methylation. Such interaction among the various components of the epigenetic machinery re-emphasizes the integrated nature of epigenetic mechanisms involved in the maintenance of global gene expression patterns.[121-124]

Although no DNA demethylase activity has been convincingly identified, several mechanisms have been proposed to account for the loss of DNA methylation - for example, DNA deaminases of the Aid/Apobec family have been shown to catalyze deamination of 5-methylcytosine resulting in T:G mismatch, which may lead to DNA demethylation if the mismatch is repaired.[75, 125]

Interestingly, a recent study has proposed that DNMTs themselves have dual roles in CpGmethylation and active demethylation of 5-methyl CpGs through deamination and recruitment of DNA glycosylase, and base excision repair proteins. Also, is thought that 5-hydroxymethylcytosine may prompt DNA demethylation. 5-methylcytosine is converted to its hydroxy state by the enzyme TET1.[126, 127]

Introduction - Epigenetics

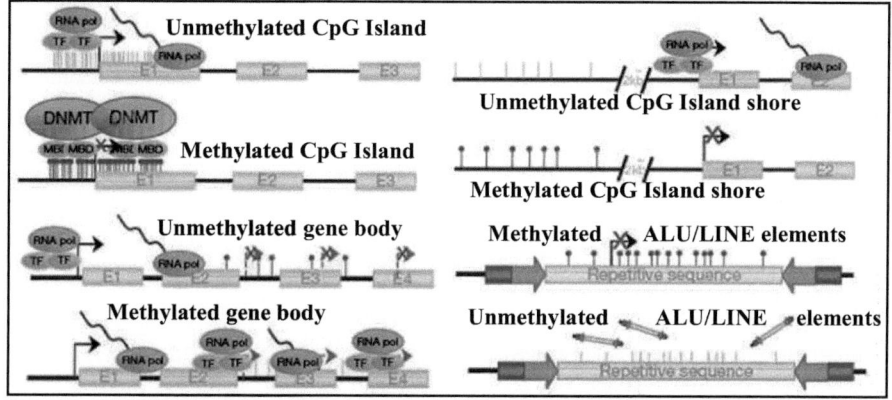

Figure 7 DNA methylation patterns. DNA methylation can occur in different regions of the genome. CpG islands at promoter regions (top right) are normally unmethylated, allowing transcription. RNA polymerases (RNA pol) and transcription factors (TF) bind to promoters and Exons (E1 and E2). The same pattern is observed in CpG island shores up to 2 kb upstream of a CpG island (top left). Aberrant hypermethylation leads to transcriptional inactivation in both cases through binding of Methyl-CpG-binding domain proteins and DNA methyltransferases (DNMT). Methylation in gene bodies (bottom left) prevents transcription initiations at incorrect sites (i.e. introns/exons). Repetitive genomic sequences (e.g. ALU and LINE elements, bottom right) are normally methylated, but if they become unmethylated retrotransposons are reactivated.[91]

Small RNAs and cancer

Micro-RNAs (miRNAs) are small non-coding RNAs (about 22 nucleotides long) that regulate gene expression through posttranscriptional silencing of target genes. These oligonucleotides are first synthesized as long, noncoding RNAs that are processed by the RNA cleaving enzyme DROSHA in the nucleus, transported into the cytoplasm in the form of short hairpin RNAs, and further cleaved by the enzyme DICER into their final configuration of double-stranded miRNAs. Sequence-specific base pairing of miRNAs with untranslated regions of target messenger RNAs (3'-UTRs) within the RNA-induced silencing complex (miRISC) results in target messenger RNA degradation or inhibition of translation. These small RNAs are expressed in a tissue-specific manner and control a wide array of biological processes including cell proliferation, apoptosis and differentiation. The list of miRNAs identified in the human genome and their potential target genes is growing rapidly, demonstrating their extensive role in maintaining global gene expression patterns. A single miRNA

can have hundreds of target mRNAs, highlighting the implication of this gene regulation system in cellular functions. [77, 87, 128]

Accumulating evidence from studies comparing miRNA expression profiles in tumors and corresponding normal tissues indicate widespread changes in miRNA expression during tumorigenesis. Micro-RNAs can function as either tumor suppressors or oncogenes depending upon their target genes. Many tumor suppressor miRNAs that target growth-promoting genes are repressed in cancer, in contrast oncogenic miRNAs, which target growth inhibitory pathways, are often upregulated in cancer. Changes in miRNA expression can be achieved through various mechanisms including chromosomal abnormalities, transcription factor binding and epigenetic alterations. Many miRNAs are controlled by DNA methylation, particularly if they are embedded in a CpG island. Since such epigenetic repression of tumor suppressor miRNAs can be potentially reversed by treatment with chromatin modifying drugs, they can serve as promising targets for epigenetic therapy. [73, 129, 130]

In addition, miRNAs can also modulate epigenetic regulatory mechanisms inside a cell by targeting enzymes responsible for DNA methylation (DNMT1, DNMT3A and DNMT3B)[124] and histone modifications (EZH2, HDAC1 and HDAC4).[131-133] Loss of the DICER complex leads to downregulation of DNA methyltransferases in mouse embryonic stem cells, which can be rescued by transfection of the miR-290 cluster, re-establishing the correct methylation patterns.[134-136]

1.3.3 Altered DNA Methylation in Cancer

Cancer initiation and progression are accompanied by profound changes in DNA methylation that were the first epigenetic alterations identified in cancer. A cancer epigenome is marked by genome-wide hypomethylation and site-specific CpG island promoter hypermethylation (see figure below). While the underlying mechanisms that initiate these global changes are still under investigation, recent studies indicate that some changes occur very early in cancer development and may contribute to cancer initiation. This recent work also suggests that the global epigenetic changes in cancer may involve the dysregulation of hundreds of genes during tumorigenesis. The mechanism by which a tumor cell accumulates such widespread epigenetic abnormalities during cancer development is still not fully understood.[73, 90]

Colon cancer has become a paradigm in epigenetic research and it is well recognized that epigenetic abnormalities arise in the earliest steps of colorectal cancer development, in fact aberrant methylation patterns have been identified in preneoplastic lesions including dysplastic aberrant crypt foci, which are considered pre- cursors of colon cancer, and in hyperplastic polyps. The first observed epigenetic abnormality in colorectal cancer, global DNA hypomethylation, plays a significant role in tumorigenesis and occurs at various genomic sequences including repetitive elements, where it leads to increased genomic instability by promoting chromosomal rearrangements, and retrotransposons where it can result in their activation and translocation to other genomic regions, thus again increasing genomic instability. In addition, DNA hypomethylation can lead to the activation of growth-promoting genes, such as S-100 and MAGE genes, and a loss of imprinting (LOI) in tumors, like IGF2 which has been linked with an increased risk of colorectal cancer. Abnormal loss of imprinting (LOI) of IGF2 is also of special relevance since it is the only alteration that has so far been associated with both cancer and normal tissue of cancer patients. Strong associations were found between LOI in peripheral blood lymphocytes and LOI in the colon, and it is associated with a fivefold greater incidence of suffering colorectal neoplasia. Thus, DNA hypomethylation leads to aberrant activation of genes and non-coding regions through a variety of mechanisms that contributes to cancer development and progression.[86, 137]

In contrast to hypomethylation, which increases genomic instability and activates proto-oncogenes, site-specific hypermethylation contributes to tumorigenesis by silencing tumor suppressor genes. These genes are involved in cellular processes such as cell cycle, cell adhesion, apoptosis and angiogenesis which are integral to cancer development and progression. In colorectal cancer, various tumor suppressor genes have been shown to undergo tumor-specific silencing by hypermethylation, including the cyclindependent kinase inhibitor 2A/B CDKN2A/B (p16/p15), the mismatch repair enzyme MLH1, E-cadherin (epithelial) CDH1, retinoic acid receptor beta RARB and O-6-methylguanine-DNA methyltransferase MGMT as well as prostaglandin-endoperoxide synthase 2 COX2.[85]

Epigenetic silencing of such tumor suppressor genes can also lead to tumor initiation by serving as the second hit in the Knudson's two-hit model (first proposed by Carl O. Nordling in 1953 and later formulated by Alfred G. Knudson in 1971) – which states that multiple "hits" are necessary to cause cancer, depending on the activation of oncogenes and deactivation of tumor suppressor genes.[137] The theory of an epigenetic field defect may be an extended form of the Knudson hypothesis suggesting that epigenetic "hits" predispose a whole area (e.g. parts of the colon) for cancer.[138]

In addition to direct inactivation of tumor suppressor genes, DNA hypermethylation can also indirectly silence additional classes of genes by silencing transcription factors (leading to inactivation of their downstream targets) and DNA repair genes, which enables cells to accumulate further genetic lesions leading to the rapid progression of cancer. Aberrant promoter

hypermethylation is already evident at aberrant crypt foci which will progress to more malignant lesions as adenomas/adenocarcinomas, where this hypermethylation is observed to increase. This aberrant methylation is thought to be responsible, at least in part, for further fostering tumour progression. While the ability of DNA hypermethylation to silence tumor suppressor genes in cancer is well established, how genes are targeted for this aberrant DNA methylation is still unclear. Tumor-specific CpG island methylation can occur through a sequence-specific instructive mechanism by which DNMTs are targeted to specific genes by their association with oncogenic transcription factors. But it remains controversial whether cancer-related methylation of certain sequences occurs due to the presence of consensus motifs recruiting DNA methylation machinery or it happens randomly.[139, 86, 137]

One possibility is that silencing specific genes by hypermethylation provides a growth advantage to cells resulting in their clonal selection and proliferation. This would suggest that certain phenotypes are under greater selective pressure than others to progress to malignant transformation. The selective advantage of these epimutations during tumor progression is possible, but it is unlikely that the multitudes of epigenetic alterations that reside in a cancer epigenome occur in a random fashion and then accumulate inside the tumor due to clonal selection. A more plausible explanation would be that the accumulation of such global epigenomic abnormalities arises from initial alterations in the central epigenetic control machinery, which occur at a very early stage of neoplastic evolution. Such initiating events can predispose tumor cells to gain further epimutations during tumor progression in a fashion similar to accumulation of the genetic alterations that occurs following defects in DNA repair machinery in cancer. The 'cancer stem cell' model suggests that the epigenetic changes, which occur in normal stem or progenitor cells, are the earliest events in cancer initiation. The idea that these initial events occur in stem cell populations is supported by the common finding that epigenetic aberrations are some of the earliest events that occur in various types of cancer and also by the discovery that normal tissues have altered progenitor cells in cancer patients. This stem cell-based cancer initiation model is consistent with the observation that tumors contain a heterogenous population of cells with diverse tumorigenic properties. Since epigenetic mechanisms are central to maintenance of stem cell identity, it is reasonable to speculate that their disruption may give rise to a high-risk aberrant progenitor cell population that can undergo transformation upon gain of subsequent genetic gatekeeper mutations. Such epigenetic disruptions can lead to an overall increase in number of progenitor cells along with an increase in their ability to maintain their stem cell state, forming a high-risk substrate population that can readily become

neoplastic on gain of additional genetic mutations. This model can also be described as 'epigenetic progenitor model', see figure below). [12, 71]

The question remains though, if the epigenetic disruptions are random by chance or follow a specific program, i.e. target specific genes. Several findings have recently emerged in support of the cancer stem cell model. Mice with a LOI at the IGF2 locus and an Apc mutation show an expansion in the progenitor cell population of the intestinal epithelium, with the epithelial cells showing higher expression of progenitor cell markers and shifting toward a less differentiated state. These mice were also at a higher risk for intestinal tumors relative to control mice. Interestingly, humans with LOI of IGF2 also show a similar dedifferentiation of normal colonic mucosa cells along with a higher risk for colorectal cancer. Also, stem cell-like characteristics of tumor cells were displayed through successful cloning of mouse melanoma and medulloblastoma nuclei to form blastocysts and chimeric mice. [72, 73, 137]

DNA methylation-induced silencing of genes involved in the regulation of stem/precursor cells' self renewal capacity is commonly observed in the early stages of colon and other cancers. Aberrant silencing of these so called 'epigenetic gatekeeper' genes in conditions of chronic stress, such as inflammation, enables stem/precursor cells to gain infinite renewal capacity thereby becoming immortal. These preinvasive immortal stem cells are selected for and then form a pool of abnormal precursor cells that can undergo further genetic mutations leading to tumorigenesis. Human ES cells with cancer cell characteristics including higher frequency of teratoma-initiating cells, growth factor and niche independence have also been found. These partially transformed stem cells display a higher expression of pluripotency markers suggesting their enhanced 'stemness' along with high proliferative capacity. Another interesting mechanism proposes a role of histone marks in the tumor-specific targeting of de novo methylation, since regions that are hypermethylated in cancer are often premarked with H3K27me3 polycomb mark in ES cells, suggesting a link between the regulation of development and tumorigenesis. This observation also partially explains the theory of "CpG island methylator phenotype" or CIMP that hypothesizes that there is coordinated methylation of a subset of CpG islands in tumors since many of these CIMP loci are known polycomb targets. This proposal emerged as a new pathway for colorectal tumorigenesis, in addition to the classic mutator or chromosomal instable (CIN) and microsatellite instable (MSI) categories, standing for a subset of sporadic colorectal tumours bearing excessive cancer-specific promoter hypermethylation. CIMP has been reported in several other tumour types such as gastric, lung, liver, ovarian and leukemias, although different sequences were analysed using different CIMP-definitions. This suggests that CIMP is not restricted to specific tumour types, but rather that concordant aberrant DNA methylation can be a general phenomenon in cancer, but involve different involve different genes. More recent studies have expanded these findings by identifying subgroups of colorectal cancers: "CIMP-I" (intermediate), "CIMP-high" and "CIMP-low" as well as CIMP-negative, each associated with different common mutations like BRAF, KRAS and TP53. While other research groups challenged the CIMP concept altogether, an independent genome-wide approach confirmed concordant methylation and therefore the existence of CIMP in colorectal cancer. In this data set, a different set of sequences (CACNA1G, IGF2, NEUROG1, RUNX3 and SOCS1) was used and it better determines CIMP and non-CIMP status than the original set of sequences. Reassuringly, this new CIMP marker set has already been successfully validated using a quantitative assay in an independent large set of colorectal tumours. Therefore, strong evidence for the existence of CIMP during tumour development exists from two independent approaches. However, CIMP is still a controversial topic that will be clarified as research on the field provides new concluding remarks to support or prove CIMP wrong.[73, 137, 140]

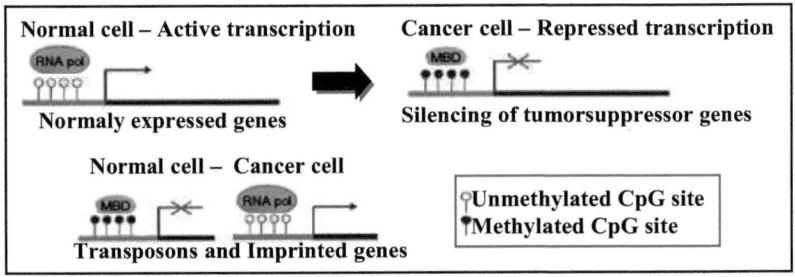

Figure 8 Changes in DNA Methylation that occur during the transformation from normal cells to in cancer cells. In normal cells, promoters of actively transcribed genes are unmethylated and can be assessed by RNA polymerases (RNA pol). Expression of other genes (e.g. viral transposons, imprinted genes) is repressed by promoter methylation and occupation of Methyl-CpG-binding domain proteins (MBD) and DNA methyltransferases. In cancer, this is deregulated, resulting in the aberrant expression of normally silent genes and repression of tumour suppressor genes.[90, 91]

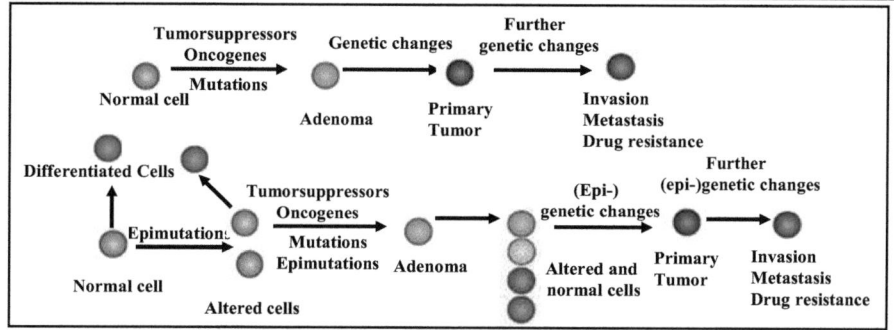

Figure 9 The classical genetic (clonal) model of cancer (top) versus the epigenetic progenitor model of cancer (bottom) as described by Feinberg.[12] In the classical model, the tumor arises through a series of mutations, activating ones in oncogenes (ONC) and silencing ones in tumour-suppressor genes (TSG). In the epigenetic model, genetic and epigenetic alterations of stem/progenitor cells within a given tissue lead to increased tumour evolution.

1.3.4 Histone modifications in cancer

Polycomb proteins, which control the silencing of developmental regulators in embryonic stem (ES) cells, provide another link between stem cell biology and cancer initiation. Polycomb proteins are commonly upregulated in various forms of cancer. In addition, genes that are marked by polycomb repressive mark H3K27me3 in ES cells are often methylated in cancer suggesting the presence of a shared regulatory framework, which connects cancer cells with stem/progenitor cell populations. Genome-wide mapping of chromatin changes occurring during tumorigenesis revealed a global loss of acetylated H4-lysine 16 (H4K16ac). Loss of histone acetylation is mediated by Histone deacetylases (HDACs) and results in gene repression. HDACs are often found overexpressed in various types of cancer and thus, have become a major target for epigenetic therapy (see below). Histone acetyltransferases (HATs), which work in concert with HDACs to maintain histone acetylation levels, can also be altered in cancer. Aberrant formation of fusion proteins through chromosomal translocations of HAT and HAT-related genes (e.g. MYST histone acetyltransferases, CREBBP and associated factors as the EP300 histone acetyltransferase) occurs for example in leukemia. In addition to changes in histone acetylation, cancer cells also display widespread changes in histone methylation patterns, such as loss of H4-lysine 20 trimethylation (H4K20me3). Furthermore, alterations in H3K9 and H3K27 methylation patterns are associated with aberrant gene silencing in various forms of cancer. Dysregulation of HMTs responsible for repressive marks results in altered distribution of these marks in cancer and leads to aberrant silencing of tumor suppressor genes, like alterations in histone acetylation (see figure below). For example, EZH2, which is the H3K27 histone methyltransferase, is overexpressed in breast and prostate cancer. Increased levels of EHMT2, the H3K9 histone methyltransferase, has been found in liver cancer and is implicated in perpetuating malignant phenotype possibly through modulation of chromatin structure. Chromosomal translocations of MLL, the H3K4 histone methyltransferase, lead to ectopic expression of various homeobox (HOX) genes and play a key role in leukemia progression.[87, 91, 141]

In addition to HMTs, lysine specific-demethylases (HDMs) that work in coordination with HMTs to maintain global histone methylation patterns are also implicated in cancer progression. KDM1, the first identified lysine demethylase, can effectively remove both activating and repressing marks (H3K4 and H3K9 methylation, respectively) depending on its specific binding partners, thus, acting as either a corepressor or a co-activator. After KDM1, several other histone lysine demethylases have been discovered, like Jumonji C domain proteins. Several of these HDMs are upregulated in prostate cancer, thus, making them potential therapeutic targets. However, since HDMs can perform both activating and repressive functions, it is essential to first understand their precise context dependent roles before their therapeutic inhibition can be used as an effective cancer treatment strategy. Despite these challenges, targeting HDMs is a promising treatment option for the future as revealed by a recent study which showed that inhibition of KDM1 in neuroblastoma causes decreased proliferation *in vitro* and inhibition of xenograft growth. [73, 142]

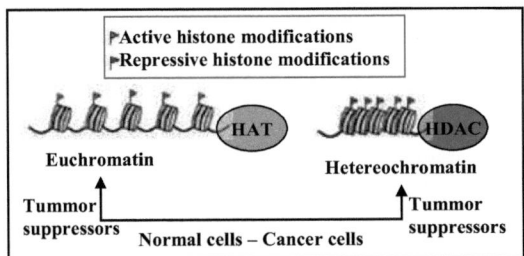

Figure 10 Histone modifications in cancer.[90] In normal cells, tumor suppressor genes have active histone modifications and attached histone acetyltransferases (HAT), forming an open chromatin formation (euchromatin). This state is altered through changes in the histone marks, in turn recruiting histone deacetylases, leading to a closed chromatin formation (heterochromatin). The inverse situation refers to oncogenes.

Introduction - Epigenetics

1.3.5 Epigenetic therapy in cancer

The reversible nature of the profound epigenetic changes that occur in cancer has led to the thinking of a possible "epigenetic therapy" as a treatment option. The aim of epigenetic therapy is to reverse the causal epigenetic aberrations that occur in cancer, leading to the restoration of a "normal epigenome". More specifically, this has resulted in an extensive search for new drugs that are capable of re-activating epigenetically silenced genes. In particular, drugs capable of reversing aberrant DNA methylation and histone acetylation patterns by inhibiting DNMTs and HDACs have been extensively explored.[73, 86, 87, 143]

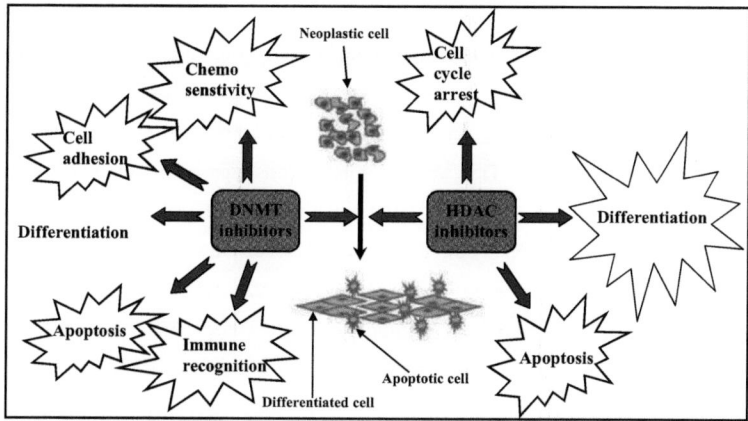

Figure 11 Epigenetically acting drugs, DNA methyltransferase (DNMT) inhibitors and histone deacetylase (HDAC) inhibitors, exert their antineoplastic effect via several mechanisms which eventually result in differentiation or cancer cell death.[87]

DNMT inhibitors

DNA methylation inhibitors were among the first epigenetic drugs proposed for use as cancer therapeutics. The remarkable discovery that treatment with cytotoxic agents, (5)-azacytidine (5-AZA) and 5-aza-2-deoxycytidine (Decitabine), lead to the inhibition of DNA methylation that induced gene expression and caused differentiation in cultured cells led to the realization of the potential use of these drugs in cancer therapy. These nucleoside analogs get incorporated into the DNA of rapidly growing tumor cells during replication and inhibit DNA methylation by trapping DNA methyltransferases onto the DNA, leading to their depletion inside the cell. This drug induced reduction of DNA methylation causes growth inhibition in cancer cells by activating tumor suppressor genes aberrantly silenced in cancer. Azacitidine and Decitabine have now been FDA

approved for use in the treatment of myelodysplastic syndromes (MDS) and promising results have also emerged from the treatment of other hematological malignancies such as acute myeloid leukemia (AML) and chronic myeloid leukemia (CML) using these drugs. However, the use of these drugs is complicated because they are chemically unstable in water, and they have been found to suppress the growth and proliferation of blood cells from the myeloid lineage, leading to toxicity problems. By contrast, other nucleoside analogues, such as 5-fluoro-2-deoxycytidine and zebularine, are much more stable in aqueous solution and less toxic compared to azacitidine. Zebularine is especially promising as a specific anticancer drug as its effects seem to be more selective for cancer cells than non-malignant cells and can be orally administered. But the ability of these drugs to be incorporated into DNA raises again concerns regarding their potential toxic effect on normal cells. However, since these drugs only act on dividing cells, one can argue that treatment with these drugs should mainly target rapidly dividing tumor cells and should have minimal effects on slowly dividing normal cells. This argument has been supported by studies demonstrating minimal side effects of long-term low doses treatment with DNA methylation inhibitors. A number of non-nucleoside analogue DNMT inhibitors have also been described, which can effectively inhibit DNA methylation without being incorporated into DNA. The development of several small molecule inhibitors such as SGI-1027 (quinoline-based compound), RG108 (N-Phthalyl-L-tryptophan) and MG98 (a 20-mer antisense compound with a phosphorothioate backbone which targets the 3'-untranslated region of the DNMT1 gene) is being actively pursued as a step in the direction of alternative approaches to cytidine analogs. These molecules can achieve their inhibitory effects by either blocking catalytic/cofactor-binding sites of DNMTs or by targeting their regulatory messenger RNA sequences; however, the weak inhibitory potential of these drugs indicates a need for the development of more potent inhibitory compounds in future. For RG108, its hydrophobicity makes it also less valuable as an anticancer drug. The demethylating potential of three other of such drugs; the major active constituent of green tea, epigallocatechin-3-gallate (EGCG), hydralazine, and procainamide have been evaluated in a study concluding that they are also weak inhibitors compared to decitabine, which is by far the most effective. The cytotoxic effects of azacitidine and decitabine are related to the formation of high levels of enzyme–DNA adducts, when used at relatively high concentrations. However, demethylation of tumor suppressor genes occurs when used at non-cytotoxic concentrations, for this reason, the effects of nucleoside analogues are now explored at lower concentrations for longer durations to favor methylation reversal over cytotoxic effects. The possible clinical use of other improved DNA methylation inhibitors is currently under investigation in Phase I to Phase III trials. The mechanisms by which the nucleoside analogues exert

their effects on the cells may be divided into those related to DNMT inhibition and those not related to demethylation of the DNA. The nucleoside analogues have been shown to be s-phase specific when used at low concentrations, and it is thought that they mainly exert their effects after incorporation into the DNA during replication in the S-phase of the cell cycle. When the DNMTs are attracted to hemimethylated DNA they will become covalently linked to the nucleoside analogues to form enzyme–DNA adducts. This results in a cellular depletion of DNMTs and subsequent hypomethylation of newly synthesized DNA strands. However, the molecular effects of the nucleoside analogues are dependent on their diverse chemical structures. Compared to cytidine, azacitidine and decitabine have nitrogen in place of carbon at position 5 in the pyrimidine ring. Zebularine does not have a nitrogen atom at this position but differs from cytosine by not having the amino group at the carbon-4 position. Finally, azacitidine is a ribonucleoside, whereas decitabine and zebularine are deoxyribonucleosides. For this reason, azacitidine also binds to RNA and thereby interrupts mRNA translation. If the nucleoside analogues exert their effects only through DNMT inhibition, it would be expected that each of the individual drugs would have similar effects on the transcriptome in a given cell line as the set of genes regulated in a methylation-dependent fashion is constant. However, it has been recently shown that the transcriptional changes in an acute myeloid leukemia cell line after individual treatment with decitabine, azacitidine or zebularine showed remarkably little overlap. Importantly, transcripts that showed a response also to treatment with the non-DNMT inhibiting cytosine analogue, cytarabine, were excluded to account for cytotoxic effects not related to DNMT inhibition. Furthermore, a considerable number of genes were down regulated after treatment with the DNMT inhibitor. This finding is inconsistent with the epigenetic paradigm that methylated genes are silenced unless the drugs have other effects apart from inhibiting the DNMTs as well. When the effects of decitabine, DNMT knockout models, and the HDAC inhibitor Trichostatin A (TSA) on gene expression were compared, a similar conclusion was reached. Since the expression profile of the decitabine treated cells resembled the profile of the TSA treated cells more closely than the DNMT knockout models, the drug does not only function by inhibiting DNMTs. Furthermore, the effects on gene expression did not seem to depend on dosage and duration, which would be expected if the drug acts on gene expression solely by incorporation into the DNA during replication in the s-phase of the cell cycle. It has been suggested that demethylation of DNA can be an active process perhaps mediated through an enzymatic protein–RNA complex, this may account for gene expression not being dependent on dosage and duration of treatment with decitabine. It has also been suggested that decitabine may directly influence the stability of methylation and chromatin marks either directly or through protein modifications, this could account for the observations that unmethylated genes become activated in response to DNMT inhibitors as gene silencing may not always be dependent on DNA methylation DNMT inhibitors may also enhance the expression of microRNAs modifying the epigenome independent on DNA methylation. [73, 86, 143-146]

Table 3 DNMT inhibitors (adapted from[86])

DNMT inhibitor	Chemical nature	Clinical status
Azacitidine	analogue of cytidine	FDA approved (MDS)
Decitabine	analogue of cytidine	FDA approved (MDS)
Zebularine	analogue of cytidine	-
5-fluoro-2'-deoxycytidine	analogue of cytidine	Phase I
epigallocatechin-3-gallate	catechin	Phases I, II
Hydralazine (Apresoline)	Non-nucleoside analogue	Phases I, II, III
RG108	Non-nucleoside analogue	-

Histone modification inhibitors

Since aberrant gene silencing in cancer is also associated with a loss of histone acetylation, re-establishing normal histone acetylation patterns through treatment with HDAC inhibitors have been shown to have antitumorigenic effects including growth arrest, apoptosis and the induction of differentiation. These antiproliferative effects of HDAC inhibitors are mediated by their ability to reactivate silenced tumor suppressor genes. Suberoylanilide hydroxamic acid (SAHA or Vorinostat), which is an HDAC inhibitor, has now been approved for use in clinic for treatment of T cell cutaneous lymphoma. Several other HDAC inhibitors such as the depsipeptides like Romidepsin as well as sodium phenylbutyrate are currently under clinical trials - in fact today, an entire array of drugs with HDAC inhibitory effects has been described and many are currently under clinical trials, however Vorinostat and Romidepsin (trade name Istodax) are so far the only HDAC inhibitors which are Food and Drug Administration (FDA) approved.[147-149]

The vast majority of HDAC inhibitors are designed to interfere with the catalytic domain of HDACs and thereby block substrate recognition and induce gene expression. Eighteen different human HDAC isoforms have been described which can be divided into four classes based on structural homologies between human and distinct yeast HDACs - Class I HDACs (HDACs 1, 2, 3, and 8) are related to the yeast RPD3 deacetylase, class IIa (HDACs 4, 5, 7, and 9) and class IIb (HDACs 6, and 10) and are homologous to the yeast Hda1 deacetylase, Class III HDACs consist of seven HDACs (SIRT1 to SIRT7) and share homologies with the yeast silent information regulator 2 (Sir2) family, class IV, only has one member, HDAC11, which shows similarities to both class I and class II HDACs.[86]

Since aberrant expression of different HDAC isoforms has been associated with different malignancies it is of interest to design isoform specific HDAC inhibitors but this remains difficult because the approximately 400 residues that comprise the catalytic domain of classes I, II, and IV HDACs are well conserved. Furthermore, Most of the described HDAC inhibitors only affect classes I and II HDACs, which are zinc-dependent. Therefore, another challenge is to design HDAC inhibitors that are unable to bind to the hundreds of zinc dependent enzymes that are involved in many different metabolic processes. The sirtuin Class III HDACs are dependent on the coenzyme nicotinamide adenine dinucleotide (NAD+) and are, therefore, inhibited by nicotinamide, as well derivatives of NAD, dihydrocoumarin, naphthopyranone, and 2-hydroxynaphaldehydes. The HDAC inhibitors described thus far vary greatly in structure and origin and they can be divided into different classes based on distinct chemical properties - short-chain fatty acids (such as phenylbutyrate and valproic acid), hydroxamic acids (such as vorinostat/SAHA, belinostat or PXD101 and panobinostat or LBH589), cyclic tetra- and depsipeptides (Romidepsin, Trapoxin B) and benzamides (e.g. entinostat or MS-275 and mocetinostat or MGCD0103). It could be expected that most HDAC inhibitors would have a global effect on gene expression as they have been found to block one or several classes of HDACs. This does not seem to be the case, however, as several microarray studies have revealed that HDAC inhibitors in general only affect a small fraction of the transcriptome, but interactions between HDACs and a large number of non-histone proteins such as transcription factors, DNA repair enzymes, chaperone and structural proteins and signal transduction mediators have been shown, establishing the role of HDACs as key-players in many different cellular processes. Therefore, the sum of the various interactions makes it difficult to establish the precise mechanism of HDACs, and in turn to develop HDAC inhibitors capable of re-activating tumor suppressor genes without undesirable effects.[73, 86, 150, 151]

Introduction - Epigenetics

Table 4 Inhibitory profile of HDAC inhibitors (adapted from[148])

	Inhibitor	class I				class IV
		HDAC1	HDAC2	HDAC3	HDAC8	HDAC11
pan-inhibitors	TSA					nd
	Vorinostat (SAHA)					nd
	NVP-LAQ824					nd
	Panbinostat					nd
	Belinostat					nd
	PCI-24781					nd
	MS-275					nd
	MGCD0103					nd
class I inhibitors	Depsipeptide			nd	nd	nd
	Apicidin					nd
	Valproic acid					nd
	Trapoxin n		nd	nd	nd	nd
	SB-429201		nd			nd
	Bispyridinum diene				nd	nd
	SHI-1:2					nd
	R306465		nd	nd		nd
	SB-379278A		nd			nd
	PCI-34051					nd
	Cpd2		nd	nd		nd
	APHA derivatives		nd	nd	nd	nd
class II inhibitors	Tubacin		nd	nd		nd
	Mercaptoacetamide					nd
	NCT-10a/14a		nd	nd	nd	nd

	Inhibitor	class IIA				class IIB	
		HDAC4	HDAC5	HDAC7	HDAC9	HDAC6	HDAC10
pan-inhibitors	TSA		nd				nd
	Vorinostat (SAHA)		nd				nd
	NVP-LAQ824		nd				nd
	Panbinostat		nd				nd
	Belinostat		nd				nd
	PCI-24781	nd	nd	nd	nd		nd
	MS-275		nd				nd
	MGCD0103				nd		nd
class I inhibitors	Depsipeptide		nd	nd	nd		nd
	Apicidin						nd
	Valproic acid	nd	nd				nd
	Trapoxin n		nd	nd	nd		nd
	SB-429201	nd	nd	nd	nd	nd	nd
	Bispyridinum diene		nd	nd	nd	nd	nd
	SHI-1:2				nd		nd
	R306465	nd	nd	nd	nd		nd
	SB-379278A	nd	nd	nd	nd	nd	nd
	PCI-34051	nd	nd	nd	nd		nd
	Cpd2	nd	nd	nd	nd		nd
	APHA derivatives		nd	nd	nd	nd	nd
class II inhibitors	Tubacin	nd	nd	nd	nd		nd
	Mercaptoacetamide	nd	nd	nd	nd		
	NCT-10a/14a		nd	nd	nd		

strong inhibition (EC50 < 5fold x EC50 relative to most sensitive HDAC isoform)
weak inhibition (EC50 > 5fold x EC50 relative to most sensitive HDAC isoform)
no inhibition (EC50 > 100fold x EC50 relative to most sensitive HDAC isoform)
nd — no data published

Interaction

The interaction between different components of the epigenetic machinery has led to the exploration of effective combinatorial cancer treatment strategies, which involve use of both DNA methylation and HDAC inhibitors together. Such combination treatment strategies have been found to be more effective than individual treatment approaches. Initially, it has been demonstrated that the administration of the HDAC inhibitor Trichostatin A alone does not re-activate densely methylated tumor suppressor genes, but when the cancer cells were treated with the DNMT inhibitor decitabine first, a synergistic effect of the two drugs could be observed - the derepression of certain putative tumor suppressor genes was only seen when azacitidine and Trichostatin A were combined. As another example, antitumorigenic effects of depsipeptide were enhanced when leukemic cells were simultaneously treated with decitabine. Synergistic activities of DNA methylation and HDAC inhibitors were also demonstrated in a study showing greater reduction of lung tumor formation in mice when treated with phenylbutyrate and a DNMT inhibitor. [73, 86, 152, 153]

Many HDAC inhibitors, including Trichostatin A, belinostat, and vorinostat have also been shown to act as synergists with a large number of conventional chemotherapeutic drugs such as paclitaxel, gemcitabine, cisplatin, etoposide and doxorubicin. Synergistic effects of decitabine in combination with paclitaxel and cisplatin have been demonstrated in various cell lines. In particular, the administration of DNMT inhibitors and/or HDAC inhibitors before chemotherapy seems to be a promising strategy to overcome the development of multidrug resistance, as acetylation of core histones provides an open chromatin configuration, making the DNA more accessible to the drugs. Pre-treatment of cancer cell lines with either Trichostatin A or vorinostat before applying chemotherapeutics such as etoposide phosphate (topoisomerase inhibitor), ellipticine (a modulator of p53, induces apoptosis), doxorubicin (an anthracycline antibiotic, works by intercalating DNA) and cisplatin increased the sensitivity of the drugs more than 10 fold in a brain tumor cell line in a cell specific manner. Applying the drugs in reverse order, initiating with the chemotherapeutic drugs did not have an effect. However, cell lines treated with a combination of sodium phenylbutyrate and valproic acid show increased expression of multidrug resistance proteins, indicating that a combination of HDAC inhibitors is not advisable; however, a combination of different classes of inhibitors might be a solution to the problem, since they target different classes of HDACs. For DNMT inhibitors, this has not been shown so far. [86, 154]

Another HDAC inhibitor, belinostat has been found to enhance the activity of carboplatin, docetaxel and paclitaxel in ovarian cancer cells, and to inhibit growth even in multidrug resistant cells. This was observed both in vitro and in vivo settings. But in a clinical Phase II trial for belinostat in patients diagnosed with relapsed malignant pleural mesothelioma, it was found that belinostat was ineffective as mono-drug and the patients presented severe side effects, severely hampering its potential clinical value. Nevertheless, belinostat is an interesting HDAC inhibitor with a solid potential to be included in combination therapy with chemotherapeutic drugs. Vorinostat also appears to synergize with many anticancer agents such as for instance imatinib (marketed by Novartis as Gleevec in the US or Glivec in Europe, a tyrosine kinase inhibitor), paclitaxel (also known as taxol, a taxane, inhibits mitosis by stabilizing microtubules) and carboplatin (an alkylating agent similar to cis- and oxaliplatin).[87, 155, 156]

The broad capacity of HDAC inhibitors for synergy with various chemotherapeutic drugs indicates that they lower the threshold for cancer cells to undergo apoptosis mediated by the drugs. It is consistent with this idea that many HDAC inhibitors have been found to decrease the levels of anti-apoptotic molecules and at the same time increase the levels of pro-apoptotic molecules. Nevertheless, further studies on the effect of HDAC inhibitors in combination with chemotherapeutic drugs are needed. HDAC inhibitors can also modulate the effects of ionizing radiation by changing gene expression, causing cell cycle arrest, growth inhibition and induce apoptosis. Likewise, HDAC inhibitors can reduce skin damage and protect from late radiation-induced effects such as fibrosis and secondary tumor formation. It has been known for decades that sodium butyrate could increase the radiosensitivity of human colon carcinoma cell lines. Likewise, Trichostatin A, valproic acid, vorinostat, entinostat, bicyclic depsipeptide and hydroxamic acid analogues have been found to enhance the sensitivity towards ionizing radiation of different cell lines. Although the effect of these compounds is not fully elucidated, the consensus is that treatment of cancer cells is initiated with HDAC inhibition prior to irradiation therapy to enhance the sensitizing effect. This is important for treatment of rectal carcinomas. While HDAC inhibition at high concentrations leads to cell cycle arrest, at lower, non-toxic dose HDAC inhibitors can still modulate the irradiation sensitivity, not by cell cycle arrest, but merely by affecting the expression of genes involved in response to DNA damage such as double stranded breaks caused by the ionizing radiation. HDAC inhibitors have also been found to suppress acute skin damage and skin fibrosis and carcinogenesis following radiotherapy, normally acute and long-term side effects of radiotherapy, by repressing inflammatory processes.[86, 157]

Apart from DNA methylation and HDAC inhibitors, HMT inhibitors have also been actively explored recently. One such inhibitor compound, 3-deazaneplanocin (DZNep), was shown to successfully induce apoptosis in cancer cells by selectively targeting polycomb repressive complex 2 proteins, which are generally overexpressed in cancer. While the specificity of DZNep was challenged in a subsequent study, these findings reinforce the potential of HMT inhibitors and the need for further development of specific histone methylation inhibitors.[73, 158]

Micro-RNAs also represent promising targets for epigenetic therapy. Downregulation of the oncogene BCL6 via reactivation of miR-127 following treatment with 5-AZA and 4-phenylbutyric acid strongly advocates in favor of the potential of a miRNA-based treatment strategy. In addition, the introduction of synthetic miRNAs, which mimic tumor suppressor miRNAs, can be used to selectively repress oncogenes in tumors. Micro-RNAs, such as miR-101 that targets EZH2, can be used to regulate the aberrant epigenetic machinery in cancer that may assist in restoring of the normal epigenome. However, the lack of efficient delivery methods is a major hurdle in the effective use of this strategy. Development of efficient vehicle molecules for targeted delivery of synthetic miRNAs to tumor cells is of prime importance in future.[73, 131, 143, 159]

1.3.6 Epigenetic biomarkers for cancer

Many genes show great promise as specific DNA methylation biomarkers for early cancer diagnostics, for predicting prognosis, and for predicting response to therapy as well as detecting disease recurrence. Epigenetic biomarkers in easy accessible body fluids such as blood, sputum, or urine that allows detection and diagnosis of tumors at an early stage would be ideal. However, in these types of samples, tumor derived material is hard to detect because of the presence of material from normal cells, and thus highly sensitive methods are needed. DNA methylation biomarkers offer several advantages over DNA mutation, mRNA or protein biomarkers. First, DNA is a stable molecule that can be easily isolated from body fluids and tissues as opposed to RNA needed for Reverse transcription polymerase chain reaction (RT-PCR) assays or proteins for Enzyme-linked immunosorbent assay (ELISA) based methods. Furthermore, DNA containing the methylation information can be isolated from formalin fixed paraffin embedded (FFPE) tissue and used in PCR based analysis.[86, 160, 161]

Second, the methylation signal to be detected is positive as opposed to loss of heterozygosity or mutations that can be hard to detect in the presence of an excess of normal DNA, which is a clear advantage over detection methods for genetic changes. Third, sample handling protocols are not as strict as those required for cDNA or protein expression analysis. The most sensitive methods for DNA methylation detection are generally based on PCR amplification of single locus biomarkers, these techniques are quantitative in nature and require only moderate sample purities (since the DNA is usually bisulfite converted and then again purified) compared to detection of genetic alterations. This takes advantage of the selective converting power of bisulfite for unmethylated cytosines to uracil, but not methylated cytosines. The detection of circulating tumor derived methylated DNA in plasma and serum has been shown to reflect methylation patterns commonly found in various types of primary tumors, thus allowing diagnosis of these and reliable detection of altered methylation patterns of cancer cell DNA in plasma, stool, sputum and urine sediments that can be easily achieved via non-invasive approach have been well received and highlighted by the published results in various types of cancer.[162, 163]

Examples for diagnostic markers include methylation of the glutathione S-transferase gene (GSTP1) in 80 to 90% of patients with prostate cancer, but it not in benign hyperplastic prostate tissue. Since Hypermethylation of CpG islands can be a marker of cancer cells in all types of biologic fluids and biopsy specimens, making detection of GSTP1 methylation in urine a possible clinical application. Another example is CDKN2A methylation as a biomarker for early detection of lung cancer in the sputum of smokers. Hypermethylation of CDKN2A has been also linked to poor outcome in colorectal cancer, similar to death-associated protein kinase (DAPK) in lung caner and epithelial membrane protein 3 (EMP3) in brain cancer, respectively. The prognostic value of DNA methylation biomarkers has also been demonstrated for a number of different markers in other cancers. Promoter methylation of CDKN2B, HIC1, CDH1 and ESR1 for instance predicts poor prognosis in early-stage patients diagnosed with myelodysplastic syndrome. [43, 164-167 161, 168]

The hypermethylation of particular genes is potentially a predictor of the response to treatment. Silencing of the MGMT gene due to methylation of its promoter region is an independent predictive biomarker of favorable outcome in glioblastoma patients treated with the alkylating agents temozolomide or carmustine (a mustard gas-related compound also known as bis-chloronitrosourea = BCNU), thus providing an example of a DNA methylation biomarker capable of predicting response to treatment. MGMT reverses the addition of alkyl groups to the guanine base of DNA and is thus a point of attack for alkylating agents. Moreover, the hypermethylation of MGMT in untreated patents with low-grade astrocytoma and other tumor types is a marker of a poor prognosis and it is probably related to the accumulation of mutations in these tumors. The potential of the methylation status of MGMT and other DNA-repair genes to predict the response to chemotherapy has also been seen with cyclophosphamide, also known as cytophosphane, a nitrogen mustard alkylating agent and MGMT in diffuse large B-cell lymphoma as well as cisplatin and MLH1 in ovarian cancer; irinotecan and the Werner Syndrome gene (WRN) in colon cancer and methotrexate (abbreviated MTX and formerly known as amethopterin, acts by inhibiting the metabolism of folic acid) and the reduced folate carrier gene (SLC19A1) in primary central nervous system lymphomas.

If DNMT inhibitors and HDAC inhibitors mainly function by reactivating essential tumor suppressor genes, DNA methylation as a biomarker may, in many situations, be capable of predicting response to treatment with these epi-drugs. However, the focus on DNA methylation as a biomarker has mainly been on early diagnostics, and so far very few studies have evaluated DNA methylation as a biomarker for response to treatment with DNMT inhibitors and/or HDAC inhibitors. Detection of low level methylation also shows great potential in the molecular monitoring of established disease after therapy in plasma and serum, e.g., CDKN2B methylation in acute myeloid leukemia. Baseline methylation status of
CDKN2B may also predict response to treatment with 5-azacitidine, this may suggest that patients with higher methylation levels may be candidates for higher doses and/or combination strategies. Although many promising DNA methylation biomarkers have been identified for diagnostic purposes, their use in clinical settings is still limited. This is often due to the lack of sufficient diagnostic specificity and sensitivity required for a diagnostic test. For this reason, panels of biomarkers may be needed in order to ensure sufficient specificity and sensitivity.[137, 139, 169]

The application of DNA-hypermethylation markers as tumor markers in routine clinical practice will require rapid, quantitative, accurate, and cost-effective techniques and objective criteria for selection of the genes that are applicable to different tumor types.

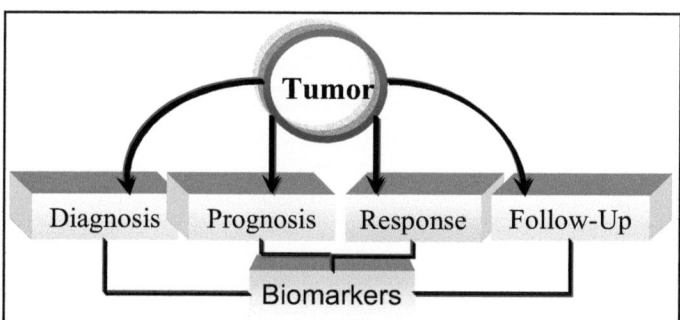

Figure 12 Possible applications of DNA hypermethylation markers (adapted from[139]).

2. Methods

2.1 Patients and tissue samples

Samples were obtained from patients undergoing CRC surgery or chemotherapy at the University Hospitals in Munich, Mannheim, Bochum, Berlin and Dresden (all Germany). Tissues were obtained during resection of the primary tumor or by biopsy. Samples were either snap frozen or formalin-fixed, paraffin-embedded (FFPE) and histology was verified by an experienced pathologist. Informed consent was obtained prior to enrollment in the study and the study was approved by the Human Subjects Committee of the University of Munich. In detail, the first 148 patient samples (74 tumor and 74 nontumor), where snap frozen samples from the University Hospital in Munich (originally collected in the University Hospital Magdeburg) which were used for methylation screening of the three selected markers TFAP2E, TUSC3, RXFP3 (see results section of this thesis). A total of 220 patient samples (tumor only) were sent from the clinics in Mannheim (42 samples), Bochum (74 samples in two batches a 24 and 50 samples) and Dresden (36 samples), as well as collected from the surgery (28 samples) and pathology (40 samples) departments of the University Hospital Munich for further validation of the methylation markers. Almost all of these samples were FFPE sections (or DNA lysates from FFPE material), but the 28 samples from the surgery department in Munich were snap frozen biopsies from colorectal tumors.

2.2 DNA and RNA extraction

The RNA and DNA extraction from cell lines or 10-25 mg of frozen tissue samples (stored at -80°C) was carried out using the RNeasy Total RNA Mini Kit or the QIAmp DNA Mini Kit (Qiagen, Hilden, Germany) respectively, using the manufacturer's instructions. For FFPE samples, the QIAamp DNA FFPE Tissue Kit was used. Briefly, tissue samples were put in liquid nitrogen and cut into small pieces with a sterile surgical blade, 20-30mg (10-15mg for RNA) of tissue were lysed completely by digestion with proteinase K and lysis buffer and the lysate was transferred into a Qia Spin Column. For FFPE samples, a number (3-5) of 10μm slides were cut from paraffin blocks and subsequently lysed and, de-crosslinked and deparaffinated. The lysate was then transferred into a Qia Spin Column. The DNA (or RNA respectively) was then bound to the silica membrane, washed and eluted from the spin columns using appropriate buffers and centrifugation steps (in case of RNA extraction, an on column DNAse I digestion step was also added according to

the manufacturer's instructions). After elution (typically in 50 to 100 µl Tris-HCL buffer, pH 8.0), the DNA or RNA was quantified using a Nanodrop 1000 photospectrometer (Peqlab Biotechnologie GmbH, Erlangen, Germany). Since the quantification of nucleic acids via a photospectrometer is based on the absorbance of UV light (at the wavelength of 260nm) in a specific pattern, it means that DNA and RNA have slightly different absorption which is it is dependent on the length and base composition of the molecules and their structure (e.g. single or double stranded helixes). To correct for this, an average extinction coefficient with units of ng-cm/ml is used in each case: for double-stranded DNA 50 ng-cm/ul and for single-stranded DNA 33 ng-cm/ul, for RNA 40 ng-cm/ul.

2.3 Bisulfite Treatment

For discrimination of the methylation status of CpGs, the DNA was treated with sodium bisulfite, which converts unmethylated cytosines to uracil, whereas 5-methylcytosines are not altered. Thus, bisulfite treatment changes the DNA sequence that depending on the methylation status of individual cytosine residues, which results in single nucleotide polymorphisms (cytosines and thymidines) after performing a polymerase chain reaction (PCR). During the four years of this study, kits from different manufacturers were used to achieve conversion, which all share the same principle. The bisulfite treatment was then done as stated in the manuals according to the manufacturer's instructions. Briefly, genomic DNA (typically 1-2µg) is incubated with a sodium bisulfite buffer for approximately 5 hours in a thermomixer or thermocycler with alternating cycles of 95°C and 55°C-60°C, thereby improving the rate of conversion by keeping strands of DNA physically separate through denaturation steps (at 95°C, sodium bisulfite chemically only works on single strands) and optimal temperatures for the sulphonation and deamination of cytosine to uracil. Afterwards, the DNA is bound again to a silica membrane and then washed and incubated with a buffer which contains a desulphonation agent, to get rid of the bisulfite ions and then eluted with the typical Tris-HCL buffer or distilled water (PCR grade), usually only 10-20µl. After elution the remaining DNA (typically up to 90% of the input DNA is degraded during the conversion process, since bisulfite treatment is quite harsh towards nucleic acids and cause strand breaks through depurination, thus explaining the low elution volume to increase the concentration), it was again quantified using a Nanodrop 1000 photospectrometer. Since bisulfite treated DNA is mostly single stranded, the extinction coefficient for single stranded DNA (33 ng-cm/ul) was used for quantification. For subsequent PCR reactions (see below), 10-20ng of the bisulfite treated DNA was

Methods

used as input from most samples. Originally, the EZ DNA Methylation and EZ DNA Methylation Gold Kits were used (Zymo Research Europe, Freiburg, Germany), next the identical MethylCode Bisulfite Conversion Kit (Invitrogen GmbH, Darmstadt, Germany) and then the EpiTect Bisulfite Kit (Qiagen, Hilden, Germany). The Imprint DNA Modification Kit (Sigma Aldrich, Steinheim, Germany) was also tested, but rejected for quality reasons (insufficient conversion rate). The rationale for using different kits was availability (in the beginning, Zymo Research was the only manufacturer with distribution in Europe), delivery time (after being on the market the Invitrogen kits had a much shorter delivery time), costs and pricing policy (e.g. Invitrogen being cheaper but offering less discounts), handling and efficiency (the Qiagen kits were easier to handle and had on average a higher DNA concentration left after bisulfite treatment). The quality of the DNA after bisulfite treatment was tested via PCRs using control primers for the human gene MLH1 and universal methylated human DNA as a standard (either ready to use from Zymo Research, Qiagen, Millipore or Chemicon or self-made with genomic DNA from SW480 cells), which is enzymatically methylated at all cytosine positions comprising CG dinucleotides by M.SssI methyltransferase isolated from a strain of *E. coli* (New England Biolabs GmbH Frankfurt am Main, Germany) and was bisulfite treated in the same manner as the DNA samples. For self-made creation of this standard DNA, the reaction conditions according to the manufacturer's instructions were used (catalog no. M0226).

2.4 Primer design

Methods for analysing bisulfite treated DNA can be generally divided into strategies based on methylation-specific PCR (MSP) and non-methylation-specific PCR conditions (see Figure below). In this thesis, methylation specific real-time PCR was performed by MethyLight technology.[170] Compared to classical MSP assays, this reduces the risk of false positives and contamination as well as handling errors, allowing a much greater sensitivity. Furthermore, it allows a greater flexibility in primer design, due to the inclusion of a TaqMan probe (see MethyLight analysis below), which provides several strategies for methylation detection, thus improving specificity. The greatest degree of methylation discriminatory capability is achieved by designing the primers and the probe to include CpG dinucleotides (to specifically amplify only the methylated sequences), therefore, for each analysed marker gene (TFAP2E, TUSC3, RXFP3) sequence specific primers and probes were designed which flank CpG Islands and contained at least 2 CpG dinucleotides each. Other primers and probe that do not cover any CpG dinucleotides serve as a control reaction for the amount of

input DNA (here the primers and probe for the ACTB gene). Primers for methylated TFAP2E were designed spanning CpGs in the second CpG Island, after pilot experiments implied the first CpG island as homogenously methylated in blood lymphocytes (data not shown) suggesting a conserved methylation pattern. Results from the MethyLight assays were confirmed using bisulfite sequencing (in selected cases from cell lines) and Methylation-sensitive high resolution melting analysis (MS-HRM). To avoid assay deviations due to limited DNA content/degradation in the clinical samples, only such samples were included in the analysis that showed a sufficient amount of DNA from the reference gene (ACTB), reflected by a Crossing point of >37 cycles (i.e. < 100pg bisulfite converted DNA, therefore ensuring a minimum of amplifiable bisulfite converted DNA in the 10ng of input DNA) in the MethyLight or HRM reaction, and value deviations of >1 cycle in the replicate measurements. Primer sequences are listed in tables S1a-c in **Appendix A**. MethyLight and MS-HRM conditions were as follows: 10 min 95°C for activation and then 50 cycles consisting of 15 sec 95°C and 30 sec 60°C and 10 sec 72°C (data acquisition at this step), for MS-HRM this was followed by 1 min 95°C, 1 min 40°C, 1sec 65°C and then continuous melting to 95°C. Typically a 20µl final reaction setup was used. All primers for expression analysis were designed to span several exons (if the gene in question consists of more than one) and to have a melting temperature which allowed an annealing temperature between 55°C and 58°C for PCR (compare the 60°C for MethyLight and HRM experiments). The primers were usually designed with the NCBI Primer Blast Tool (http://www.ncbi.nlm.nih.gov/tools/primer-blast/), which uses the Primer 3 Software (Copyright Whitehead Institute for Biomedical Research, Cambridge, MA, USA) and blasts the designed primers automatically against unintended priming on the whole human genome and transcriptome. In some cases, primer sequences were taken from the literature (as for the MethyLight input control ACTB primers), designed by hand (as for some bisulfite sequencing primers) or with Primer 3 only, or were designed by Epigenomics AG, Berlin, Germany. In these cases, the primer sequences were blasted against human genomic sequences and transcripts using the nucleotide blast Alignment Search Tool (http://blast.ncbi.nlm.nih.gov/) of the National Center for Biotechnology Information (NCBI, Bethesda, Maryland) website. All primers were ordered from Eurofins MWG Operon, Ebersberg, Germany.

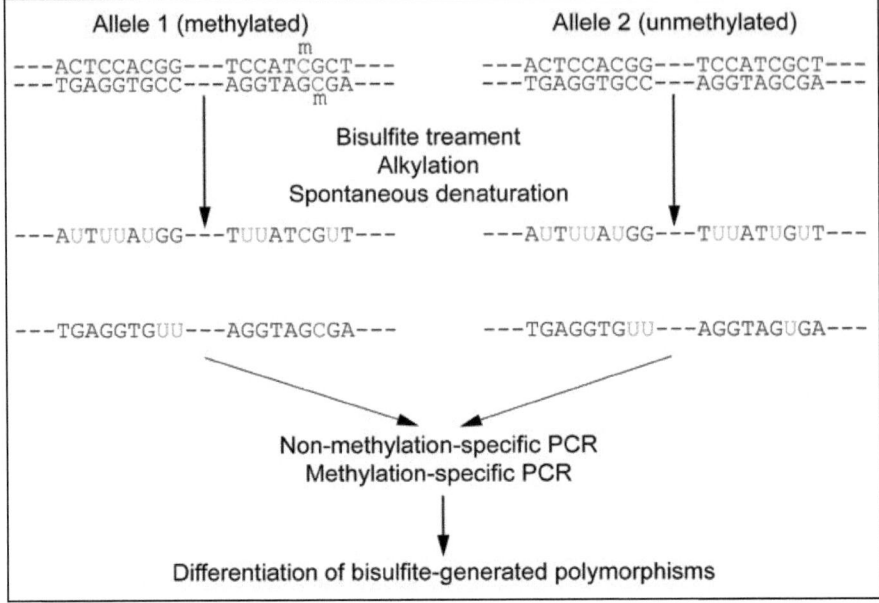

Figure 13 Overview over bisulfite conversion and subsequent amplification.

2.5 MethyLight/MS-HRM analysis

To determine the methylation status of the human tissue samples, a relative quantification analysis was done by fluorescence based real-time PCR using MethyLight technique[171] on a LightCycler 480 machine as well as the 1.5 LC480 software (Roche, Penzberg, Germany). This is essentially a fluorescent reporter probe based method - a sequence specific probe labeled with a 3´ fluorescent quencher (this is a substance that absorbs excitation energy from a fluorophore and dissipates the energy as heat, when close together with the fluorophore, it suppresses the light emission from it) dye (BHQ1 = Black hole quencher one, which quenches the whole visible spectrum) and a 5´ fluorescent reporter dye (FAM = 6-carboxyfluorescein, which emits light at 520nm producing a green color quite similar to SYBR Green I) hybridizes to the previously bisulfite converted DNA and is cleaved by the 5´ nuclease activity of the DNA polymerase during the extension phase of the PCR, which results in the separation of quencher and reporter dye and a fluorescence signal is emitted (see figure below).

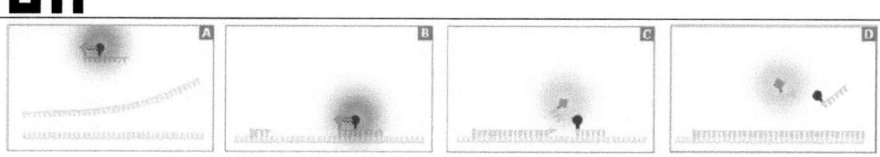

Figure 14 Principle of relative quantification by real time PCR using sequence specific probes (adapted from the LightCycler 480 Manual).

The relative quantification analysis is then based on he fact that fluorescence intensity correlates directly proportional to the amount of PCR product. Relative concentrations of DNA are determined by plotting fluorescence against cycle number on a logarithmic scale. The cycle at which the fluorescence from a sample crosses the threshold for detection of fluorescence is called the cycle threshold, Ct or crossing point, Cp. Amounts of the DNA products are then determined by comparing the results to a standard curve produced by real-time PCR of serial dilutions (typically, undiluted, 1:4, 1:16, 1:64, 1:256) of a known amount (between 10ng and 20ng) of the fully methylated standard DNA. The measured amount of methylated DNA from the gene of interest is divided by the amount of DNA from a unmethylated reference gene (typically ACTB) measured in the same sample to normalize for possible variation in the total amount (input) and quality of DNA between different samples. The concentration is expressed as a ratio of target to reference gene in the same sample, rather than an absolute value. This ratio is then normalized using the undiluted fully methylated DNA standard as a calibrator and multiplying the ratio with 100 to get a percent value, the so called percent methylation ratio (PMR). Using those PMRs from cohorts of patients, cutoff values giving the best discrimination between non-neoplastic mucosa and tumor samples were determined by receiver operating characteristic (ROC) analysis (see statistics) for each target gene. For FFPE samples, a further real-time PCR-based method to differentiate converted from unconverted bisulfite-treated DNA was used - high-resolution melting analysis (HRM). The PCR amplicons are analyzed directly by temperature ramping and resulting liberation of an intercalating fluorescent dye (a dye which shows differential fluorescence emission dependent on their association with double-stranded or single-stranded DNA like SYBR Green I but can be used a saturating concentrations because it does not inhibit PCR then, e.g. EvaGreen) during melting. The degree of methylation, as represented by the C-to-T content in the amplicon, determines the rapidity of melting and consequent release of the dye. This method allows direct quantitation in a single-tube assay, but assesses methylation in the amplified region as a whole rather than at specific CpG sites. In a HRM assay, the measured fluorescence is initially high when both DNA strands of the

Methods

amplicons are together, but diminishes as they dissociate when the temperature is raised. For each sample, this gives a characteristic melting curve which is then compared to those of dilutions of fully methylated DNA standards (typically undiluted, 1:2, 1:3; 1:4, 1:10) by normalization and temperature shifting and then using a difference plot (see figure below). The reason for choosing HRM over MethyLight alone was due to the higher sensitivity and specificity, in particular, for sensitive detection of low-level methylation and that it allows to exclude some false-positive or - negative results when compared with MethyLight data from the same sample. DNA extracted from FFPE samples offers certain challenges for PCR based methods, since even after reversal of formaldehyde modifications most of the DNA is still crosslinked to proteins and inhibitors might be still present in the lysate. This is especially true, as in our case, the exact contents and concentrations of the solutions used in the fixation and embedding process are not known (e.g. they may vary according to the manufacturer's protocol of the used embedding machine), since the formalin fixation and paraffin embedding process was not done in our lab. To check for reproducibility, MethyLight and HRM assays were tested on the 28 tumor biopsies (snap frozen material) from the surgery department of the University Hospital Munich and provided the same results in over 90% of the cases. This was only partly true for FFPE samples, as the same test gave more mixed results – while the 40 samples from Munich gave comparable results, in the 42 samples from Mannheim and the 74 samples from Bochum methylation levels could only detected with the HRM method. On average, in the case of DNA extracted from FFPE material, an approximately 10fold lower amount of amplifiable DNA could estimated compared with fresh frozen material, leading to much higher Cp values and making the use of the real-time PCR data to quantify methylation difficult (since the amount of methylated DNA is then underestimated). Using more than 10-20ng bisulfite treated DNA as starting material did not alter the results (possibly due to either left over inhibitors from the formalin and paraffin treatment or incomplete bisulfite conversion process due to crosslinking and inaccessibility of the DNA). The use of Methylation-Sensitive HRM (in this case using primers containing CpGs, since the same primers as for the MethyLight assays were used) allowed for the removal of false negatives since estimation of methylation levels in MS-HRM is performed on the basis of a comparison of melting profiles of the samples and standards of methylated DNA, thus allowing methylation discrimination based on a smaller amount of molecules then needed for MethyLight (while for MethyLight only the CpGs of the primers and probes are detected, as for HRM this is true for all CpGs in the whole amplicon).

In sum, MS-HRM[172] tends to be more sensitive then classical MSP or MethyLight[161], so MS-HRM (with the same primers as used for MethyLight) was used for all FFPE samples whenever possible.

All MethyLight assays were done in triplicates (technical replicates) for each patient sample (tumor and nontumor). The HRM assays were done without using replicates but repeated at least two times with the same conditions and lysates from the same bisulfite conversion. For both MethyLight and HRM experiments, different Mastermixes from different suppliers were used, most frequently (in this order) the LightCycler 480 Probes Master (Roche Applied Sciences, Penzberg, Germany), the EpiTect MethyLight PCR Kit and the QuantiTect Probe PCR Kit for MethyLight assays and the LightCycler 480 High Resolution Melting Master (Roche, containing the ResoLight dye), the EpiTect HRM PCR Kit (Qiagen, Hilden, Germany, using EvaGreen) as well as the MeltDoctor HRM Master Mix (Invitrogen GmbH, Darmstadt, Germany) for HRM assays. This was done to test for the robustness of the assays. When switching to a different mastermix then the one used before, a batch of 10-12 samples was measured with both mixes, to ensure the validity of the results.

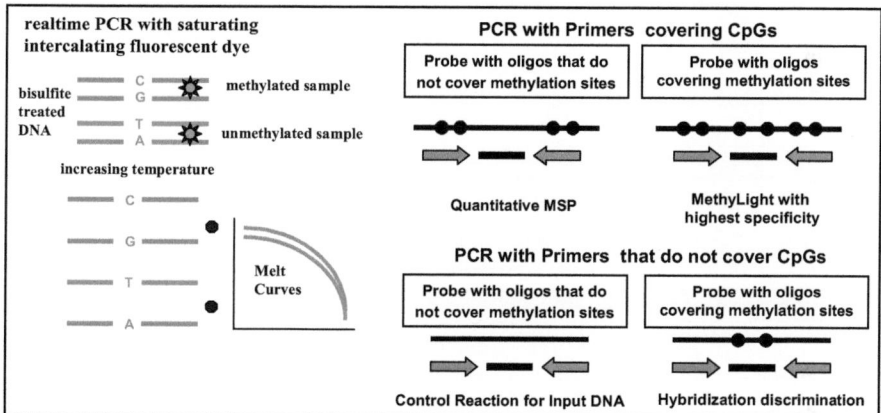

Figure 15 Overview over HRM and MethyLight technology.[173]

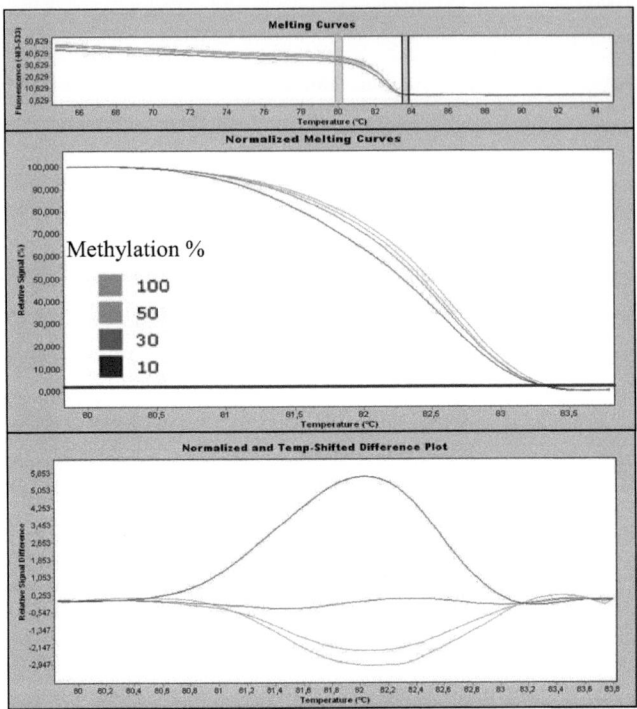

Figure 16 Examplary fluorescent data from the melting standards for the TFAP2E HRM assay using the Gene Scanning module of the LightCycler 480 Software 1.5. The melting curve data (here different dilutions of fully methylated DNA standard) is normalized by adjusting the temperature range for the fluorescent signals (top picture, here a setting for 80°C pre-melt and 84°C post-melt signals) and then temperature shifted at the point were the entire double-stranded DNA is completely denatured (around 5% of the data, red line in the middle) to easily distinguish the degree of methylated DNA by the different shapes of the melting curves. In the final step (bottom picture), the differences are further visualized by selecting a base curve (in this case, the degree of 30% methylation) from which the other curves are subtracted, thus generating a Difference Plot, which helps to cluster samples into groups with similar melting curves (i.e. those with a similar degree of methylation).

Methods

2.6 Reverse transcription polymerase chain reaction (RT-PCR)

To perform polymerase chain reaction (PCR) for gene expression, a conversion of the previously extracted RNA into cDNA is necessary. For this study expression analysis of the mRNA was performed by RT-PCR with the Verso cDNA Kit (Thermo Fisher, Darmstadt, Germany) according to the manufacturer's instructions.

Generally, the reverse transcription reaction generally includes 3 steps. First of all, primer oligonucleotides anneal to the mRNA at the 3´OH end. From this starting point on, a complementary DNA strand is synthesized with dNTPs by reverse transcriptase. After the cDNA strand is completely synthesized, the mRNA strand is degenerated due to the RNase H activity of the reverse transcriptase. The total yield of complementary DNA is dependent on the primers used for the mRNA and the efficiency of the used reverse transcriptase enzyme and the buffer conditions. Typically, either random hexamers or oligo dTs are used for RNA priming (see figure below). The combination of these two priming strategies increases sensitivity and also gives greater coverage and higher yield of cDNA from RNA. An RNase inhibitor to reduce the possibility of RNA degradation is included within the Verso enzyme mix as well. The here used cDNA Kit also contains another enzyme, the RT Enhancer, that degrades any contaminating DNA during the RT step, saving time and effort. The optimized buffer reduces RNase H activity, allowing the Verso enzyme to generate full length cDNA up to 12kb. Verso cDNA kits include. For this thesis, 1µg of total RNA was added as a template in the reverse transcription reaction (a 20µl final reaction setup). The used conditions were as follows: 2 min at 25°C and 60 min at 42°C with 2 min 95°C.

Following reverse transcription the cDNA was stored in -20°C and 1µl was used as undiluted template for PCR to determinate gene expression (see primer lists S1). For PCR, the GoTaq Green Mastermix (Promega GmbH, Mannheim, Germany), a premixed ready to use solution containing *Taq* polymerase, dNTPs, magnesium chloride and reaction buffers at optimal concentrations was used together with gene specific primers. In case of real time PCR, SYBR Green I based chemistry was used, mainly the LightCycler 480 SYBR Green I Master (Roche, Penzberg, Germany). In contrast to probe based chemistry, only primers are used, as the mastermix contains SYBR Green I as a fluorescent dye which binds to all double-stranded DNA molecules. An increase in DNA product during PCR therefore leads to an increase in fluorescence intensity and is measured at each cycle, thus allowing DNA concentrations to be quantified. However, SYBR Green I will bind to all dsDNA PCR products, including nonspecific PCR products (such as Primer dimers). This can potentially interfere with accurate quantification of the intended target sequence, but using SYBR Green I instead of a fluorescent reporter probe method is more cost effective, since no labeled oligos (probes) are needed, thus allowing the quantification of multiple target genes with regular primers (on the other hand, multiplexing, i.e. several targets per well are impossible with this chemistry).

Conditions were usually as follows: a 2 min at 95°C initial denaturation step (5-10 min for SYBR green mixes), followed by 25 to 45 cycles of 10 sec 95°C, 10-30 sec 55-60°C, 10-30 sec 72°C depending on the amplicon length and a final extension step for 5 min at 72°C.

PCR products were visualized with a 2% agarose gel with TAE buffer stained with Ethidium bromide and photographed on a gel documentation chamber (BioRad). In case of real-time PCR, the LightCycler Software 1.5 was used for relative quantification analysis (ACTB was used as reference gene).

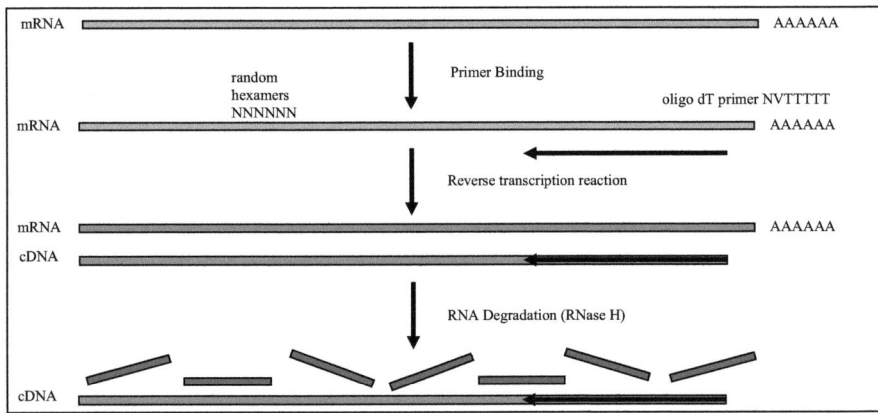

Figure 17 Principle of reverse transcription.

2.7 Cell culture and 5-aza-cytidine treatment

Colon cancer cell lines LOVO and DLD-1 were obtained from the *Deutsche Sammlung von Mikroorganismen und Zellkulturen GmbH* (Braunschweig, Germany). SW480, HT-29, HCT-116 and CACO-2 colon cancer cell lines were obtained from ATCC trough LGC standards (LGC Standards, Wesel, Germany and ATCC, Washington, DC, USA). Cells were cultured (in tissue culture flasks or tissue culture plates of different sizes) in 90% DMEM medium supplemented with 10% fetal calf serum and 2 mM L-glutamine (Invitrogen, Karslruhe, Germany) at 37°C and 5% CO_2 in a humidified incubator. CACO-2 cells were maintained in 80% DMEM with 20% FCS and 1% non-essential amino acids. By reaching confluency, cells were washed with PBS, trypsinized and

different amounts of cells, depending on the size of the flask or plate, the total number of cells and the splitting ratio, were transferred into a new culture flask and resuspended in fresh medium. For treatment with 5-aza-(2-deoxy)-cytidine, cells were seeded at a density of 1×10^6 cells/60-mm dish, after twenty-four hours incubated with azacytidine/decitabine (final concentrations 10 µmol/L; Sigma) or the same amount of DMSO as control. Culture medium was changed daily for 3 days with recurrent 5-aza-(2-deoxy)-cytidine additions. After 3 days, the cells were harvested for total RNA extraction with the RNeasy Total RNA Mini Kit (Qiagen, Hilden, Germany) and DNA extraction with the DNeasy Blood and Tissue Kit (Qiagen).

2.8 Reporter and expression vectors and subcloning

Full length TFAP2E, TUSC3 and RXFP3 coding sequences (CDS) were amplified from cDNA of SW480 cells using the GoTaq Green Mastermix (Promega) or the JumpStart REDTaq ReadyMix (also a premixed ready to use solution, Sigma, Munich, Germany) and cloned with and without FLAG epitope (e.g. pTFAP2E and pTFAP2eFlag, pTUSC3, pRXFP3, pRXFP3FTS) into the pTARGET vector (Promega, Mannheim, Germany) according to the manufacturer's instructions. Briefly, PCR amplicons were purified (using the Qiaquick PCR purification kit, Qiagen, or the PureLink Quick Gel Extraction and PCR Purification Combo Kit, Invitrogen) and ligated into the pTARGET (predigested, see Appendix E) vector using T4 DNA Ligase (Promega) and a vector:insert ratio of 3:1. Afterwards, the generated vectors were transformed into JM109 competent *E.coli* cells, (Promega) using heat shock (42°C for 45s), which were then grown in SOC medium (Sigma) for one hour and then plated onto LB/ampicillin/IPTG/X-Gal plates (self-made, all chemicals were obtained from Sigma) and incubated overnight (16-18 hours). Colonies were then selected through blue white screening and the cloned sequences were verified through amplifying and sequencing individual colonies (colony PCR). Positive colonies were then enriched and plasmids purified through a plasmid mini preparation using the PureYield Plasmid Mini Prep system (Promega). For generating FLAG tags, the full length CDS were again amplified after cloning into the pTARGET vector using adaptor primers encoding for FLAG and containing an in frame STOP codon (primers were designed to replace the native STOP codon). The DKK4 promoter sequence was also amplified from SW480 cells and cloned into the pGL3 basic vector (Promega) - pGL3-DKK4-1kb (insert size: 1kb, from minus 1kb of the transcription start site). The pRL-TK reporter plasmid (Promega) was used as an internal control reporter vector. Full length DKK4-CDS (pcDNA3-DKK4) and a DDK4 promoter plasmid - pGL3-DKK4-2kb (pGL3 basic

Methods

vector – insert: 2kb, minus 2000 bp of the transcription start) were obtained from Dr. Kolligs' lab, München[174]. To generate the pcDNA3 DKK4 vector, they used the same source (SW480 cells) to clone human DKK4 cDNA. The resulting PCR product was digested and cloned into the expression vector pcDNA3 (Invitrogen) digested with the same restriction enzymes. The sequence of this plasmid was verified by sequencing. Primer sequences are listed in **tables S1a-c** and the full sequence plus an overview is given in **Appendix E** for each insert. For subcloning of PCR products (e.g. for verification through sequencing) the TOPO TA Cloning Kit was used (Invitrogen) which contains the pCR2.1 vector (a vector which has the topoisomerase enzyme covalently attached to both of its strands' free 3' ends, so no ligase is needed) together with competent cells (DH5 alpha or TOP10).

2.9 Generating of clones with stable overexpression

SW480 clones stably overexpressing TFAP2E, TUSC3, RXFP3 and control clones using an empty pTarget vector were obtained after transfection with (e.g.pTFAP2E, pTFAP2eFlag or pTUSC3, pRXFP3) using Lipofectamine 2000 (Invitrogen, Karlsruhe, Germany), a lipid-based transfection reagent, according to the manufacturer's instructions. An empty pTargeT vector was transfected separately as control. Stable transfectants were selected upon Geneticin (G418) treatment (Invitrogen, Karlsruhe, Germany) for 2 weeks. Single colonies were picked out and transferred to either a 96 or 24 tissue culture plate and cultivated further in selective media. TFAP2E expression was assessed by quantitative RT-PCR. This transfection technology (also called lipofection) functions through the principle that plasmid DNA adheres to the surface of small liposomes (membrane-bounded bodies) due to ionical interactions and becomes included in them. Since liposomes are in some ways similar to the structure of a cell and can actually fuse with the cell membrane, this releases the DNA into the cell. The main advantages of lipofection are its high efficiency, its ability to transfect all types of nucleic acids in a wide range of cell types, its ease of use, reproducibility, and low toxicity. In addition, this method is suitable for all transfection applications (i.e. transient, stable, co-transfection, batch transfections). The plasmid DNA/liposomes ratio is chosen in excess of liposomes so that a positive charge remains and that plasmid DNA is completely complexed with the liposomes. This charge probably enables the complex to adhere to negatively charged residues of sialic acid on the cell surface following internalization.

2.10 Immunofluorescence and Immunoblotting

For verification of protein expression of the generated stable clones, immunofluorescence and protein immunoblotting were done according to the manufacturer's instructions (Abcam plc. Cambridge, UK) using polyclonal antibodies against TFAP2E and TUSC3 (in the case of RXFP3, the verification was done using the same anti-FLAG antibody which was used for chromatin immunoprecipitation). Briefly, protein immunoblotting is a technique to detect proteins with specific antibodies proteins that have been denatured and separated from one another according to their size by gel electrophoresis (using a polyacrylamide gel). The proteins are then transferred to a membrane (the blot), usually made of nitrocellulose. The gel is placed next to the membrane and application of an electrical current induces the proteins in the gel to move to the membrane where they adhere. The membrane is then a replica of the gel's protein pattern, and is subsequently stained with an antibody. Cells are usually first lysed and denaturated through heating (using buffers containing a detergent like Sodium dodecyl sulfate (SDS) and boiling at 90°C for a few minutes) and then loaded with a sample buffer on a sodium dodecyl sulfate polyacrylamide gel for electrophoresis (SDS-PAGE), where the percentage (and so thickness) of the polyacrylamide depends an the size of the protein, and run in a gel chamber filled with a electrophoresis buffer and a constant voltage for several hours. Sampled proteins become covered in the negatively charged SDS and move to the positively charged electrode through the acrylamide mesh of the gel. Smaller proteins migrate faster through this mesh and the proteins are thus separated according to size (usually measured in kilodaltons, kDa). The gel is then packed into a stack of filter papers and sponges together with the membrane und put together with a transfer buffer back into a gel chamber and an electric current is used to pull proteins from the gel into the nitrocellulose membrane. The proteins move from within the gel onto the membrane while maintaining the organization they had within the gel. As a result of this "blotting" process, the proteins are exposed on a thin surface layer for detection. In the next steps, the membrane is washed and placed into blocking buffer (usually containing 5% non-fat dry milk or 1% BSA and a minute percentage of detergent such as Tween 20 or Triton X-100) to prevent non-specific binding of the antibodies and then incubated with the primary and secondary antibodies (with washing steps in between) for any time between 30 minutes and overnight (depending on the antibody). The primary antibody recognizes the target protein, while the secondary antibody recognises the primary antibody and binds to it. For detection, the secondary antibody is linked to a reporter enzyme, which when exposed to an appropriate substrate drives a colorimetric reaction and produces a color. Most commonly, a horseradish peroxidase-linked secondary is used to cleave a chemiluminescent agent, and the reaction product produces

Methods

luminescence in proportion to the amount of protein. A sensitive sheet of photographic film is placed against the membrane, and exposure to the light (between 1 minute and 30 minutes) from the reaction creates an image of the antibodies bound to the blot. Thus, a positive band of the right size (in kDa, compared to a molecular weight marker) confirms the expression of the target protein (and possible relative quantification - e.g. if strong or weak expression) in the cells of the sample. To detect the localisation of the expressed protein in the cell, a widely used technique is light microscopy with a fluorescence microscope - immunofluorescence, a specific example of immunohistochemistry or in this case immunocytochemistry (since no successful immunostaining of tissue sections or cells was performed for this thesis, this topic is not further elaborated here, but the protocols for both immunocytochemistry and immuofluorescence are almost exactly the same). This technique uses specific antibodies labeled with fluorescent dyes to target specific proteins (for this thesis, TFAP2E and TUSC3 proteins were the targets) within a cell, and therefore allows visualisation of the distribution of the target molecule through the sample. Briefly, the cells are cultured in a plastic chamber on a glass slide (BD Falcon BioCoat culture slides, BD Biosciences, Heidelberg, Germany) and fixed with 4% paraformaldehyde, the cell membranes are then permeabilized with either acetone or methanol or 0.25% Triton X-100. Afterwards, the cells are incubated with 1% BSA in PBS with Tween 20 for blocking non-specific antibody binding (alternatively, 1% gelating or 10% serum from the species the secondary antibody was raised in can be used) and then incubated with the primary and secondary antibodies (similar to immunoblotting), typically overnight at 4°C for the primary and 1 hour at room temperature in the dark for the secondary antibody. Instead of a reporter enzyme, the secondary antibody is s chemically linked to a fluorophore (this saves the cost of modifying the primary antibodies to directly carry a fluorophore and several different primary antibodies may therefore be recognized by a single secondary antibody). For counterstaining, non-antibody methods of fluorescent staining can be used, for example, the use of DAPI (4',6-diamidino-2-phenylindole is a fluorescent stain that binds strongly to A-T rich regions in DNA) or Hoechst stains to visualize nuclei and mitochondria. The cells are then photographed under a microscope at the appropriate wavelengths for the secondary antibody (depending on the excitation and emission spectra), for this study, the Alexa Flour 488 dye was used, which emits a cyan-green color.

Methods

2.11 Transient transfections and luciferase assays

To assess the transcriptional activity in cells that were transfected with the CDS of TFAP2E, luciferase reactions were performed. In such a reaction light is emitted when luciferase acts on the appropriate luciferin substrate, which can be can be detected a by light sensitive apparatus such as a luminometer. For this purpose, the pGL3 basic luciferase reporter vector (Promega), which contains a modified coding region for firefly (*Photinus pyralis*) luciferase, but no promoter region (hence no constitutive expression) was used together with the pRL-TK vector (a wildtype *Renilla* luciferase (*Rluc*) control reporter vector) which provides constitutive expression of *Renilla* luciferase. When a promoter region containing a binding site is inserted into the pGL3 vector, the expression of firefly luciferase can be measured depending on the appropriate interaction partners (transcription factors and cofactors) present in the cell into which the vector with insert is transfected. A cotransfection with a specific regulating factor can then determine the interaction (e.g. if a certain transcription factor activates or represses a target gene by binding to the promoter region of interest) between the two genes. For this thesis, SW480, CACO-2 and HT-29 cells were cotransfected with either pGL3-DKK4-1kb or pGL3-DKK4-2kb and pRL-TK. Expression of *Renilla* luciferase provides an internal control value to which expression of the experimental firefly luciferase reporter gene is normalized. Transfections were carried out with Satisfection Transfection Reagent (Agilent, Waldbronn, Germany). This is essentially a cationic polymer, where the negatively charged DNA binds to and the complex is taken up by the cell via endocytosis, similar to the more popular Fugene (Roche, Penzberg, Germany). These kinds of reagents are less toxic to the cells than lipid based reagents. Cells were also transfected with pTFAP2E plus pGL3-DKK4-1kb or pGL3-DKK4-2kb and pRL-TK. After 3 days, cells were harvested and firefly and renilla luciferase activities (both renilla and firefly) were measured on a luminometer (displayed in relative light units RLU) using the Luciferase Dual Reporter Assay (Promega, Mannheim, Germany). The RLU ratios (renilla/firefly) were then displayed as foldchange for cells transfected with and without pTFAP2E (see results section). All experiments were repeated independently 3 times and all cells were seeded in triplicate on the 96 well plates.

2.12 Chromatin immunoprecipitation

To investigate the interaction between TFAP2E protein and the promoter of DKK4, pTFAP2eFlag SW480 clones were used for Chromatin Immunoprecipitation (ChIP) according to the protocol of ChIP assay kit (Upstate Biotechnology Inc, New York, USA) with anti-FLAG antibody (Sigma,

Methods

Munich, Germany). Primer sequences of the DKK4 promoter region flanking putative AP-2 protein binding sites are given in tables S1a-c. Briefly, the method is as follows: protein and associated chromatin in a cell lysate are reversibly cross-linked (using a formaldehyde buffer solution), then sheared by sonication, providing fragments of 300-1000bp in length. The DNA-protein complexes are selectively immunoprecipitated using specific antibodies to the protein of interest, in this case, using the FLAG antibody against the tagged TFAP2E protein and coupling the antibody to magnetic beads. After washing to remove non-specifically bound chromatin, the protein-DNA cross-link is reversed and proteins are removed by proteinase K digestion. The associated DNA fragments are then purified and identified by PCR using primers against the region of interest (i.e. in this case flanking the DKK4 promoter regions with AP-2 binding sites) or their sequence is determined by other means (e.g. molecular cloning and sequencing or microarrays).

2.13 Expression Microarray and verification of target candidates

To determinate downstream interaction partners of the selected methylation markers, an expression microarray analysis was performed. Briefly, mRNA from pTFAP2E, TUSC3 and RXFP3 SW480 clones and SW480 pTarget clones as controls were used for global expression analysis using a Human Gene 1.0 ST Expression Array (Affymetrix, High Wycombe, United Kingdom) according to the manufacturer's protocol. Genes with a more than 3-fold expression change were verified by quantitative RT-PCR (gene list see tables S1a-c). The Affymetrix Arrays (also known as a known as an *Affy chip*) are glass slides containing series of thousands of microscopic spots of DNA oligonucleotides (probes) to which cDNA or cRNA from a sample is hybridized. The probes are short sequences designed to match parts of the sequence of known or predicted mRNAs. A higher number of complementary base pairs in a nucleotide sequence lead to tighter non-covalent bonding between the two strands. After washing off of non-specific bonding sequences, only strongly paired strands will remain hybridized. The sample cDNA is fluorescence labeled, so sequences that bind to a probe sequence generate a signal that depends on the strength of the hybridization determined by the number of paired bases, the hybridization conditions (such as temperature), and washing after hybridization. Thus, Affymetrix arrays are *single-channel microarrays* or *one-color microarrays* (this means one array per sample as opposed to two color microarrays of other manufacturers), which provide intensity data for each probe or probe set indicating a relative level of hybridization with the labeled sample. However, they do not truly indicate abundance levels of a gene but rather relative abundance when compared to other samples or conditions when processed in the same

experiment (i.e. generated clones versus empty pTarget Vector controls). After normalization, the data is provided as differences in fold expression between the samples in an Excel file (see **Appendix A**).

2.14 Stress resistance and cell survival assays after drug exposure

Cells were seeded in 96 well plates (20.000 per well) and cell proliferation was measured by MTT (Sigma, Munich, Germany) and BrDU (Roche, Penzberg, Germany) assays according to the manufacturer's protocol. Briefly, a MTT assay is a simple method of measuring the activity of living cells via mitochondrial dehydrogenase activity. The main component is 3-[4,5-dimethylthiazol-2-yl]-2,5-diphenyl tetrazolium bromide or MTT, which is yellow if solubilized in tissue culture media. Mitochondrial dehydrogenases of viable cells cleave the tetrazolium ring, yielding purple MTT formazan crystals which are insoluble in aqueous solutions but can be dissolved in acidified isopropanol or hydrochloric acid with a detergent (like sodium dodecyl sulfate). The resulting purple solution is spectrophotometrically measured (by measuring at a certain wavelength, usually between 500 and 600 nm with a spectrophotometer). An increase in cell number results in an increase in the amount of MTT formazan formed and an increase in absorbance. When the amount of purple formazan produced by cells treated with an agent is compared with the amount of formazan produced by untreated control cells, the effectiveness of the agent in causing death, or changing metabolism of cells, can be deduced through the production of a dose-response curve (see results section). Changes in metabolic activity can give large changes in MTT results while the number of viable cells is constant. Therefore, as a more accurate detection method for living and proliferating cells, BrdU was used - it incorporates into the newly synthesized DNA of replicating cells (during the S phase of the cell cycle), substituting for thymidine during DNA replication. An antibody specific for BrdU is then be used to detect the incorporated chemical, thus indicating cells that were actively replicating their DNA. By using a labeled antibody, the number amount of proliferating cells ca be measured when incubated with a substrate and then quantifying the reaction product by either measuring the light emission using a scanning multi-well luminometer (luminescence ELISA reader) or by measuring the absorbance using a scanning multi-well spectrophotometer (ELISA reader). For assessment of stress induced apoptosis, cells were seeded in 96 well plates (50.000 per well) and treated with TNF-alpha (100 ng/µl) and chlorhexamide (100 µg/µl) for 4 days. Resistance of cells to chemotherapeutic drugs was investigated after treatment with 5-fluorouracil (50 µg/ml or 38 µM), oxaliplatin (60 µM or 24 µg/ml) or irinotecan (20 µM or 13 µg/ml) and proceeding with an MTT assay.

2.15 Invasion and Adhesion

To study cell invasion and adhesion, BioCoat Matrigel Invasion Chambers (BD Bioscienes) which provide cells with the conditions that allow assessment of their invasive property in vitro were used. These chambers are inserts for standard 24 well tissue culture plates and must be rehydrated before use. They contain a membrane with a thin layer of MATRIGEL Basement Membrane Matrix which serves as a reconstituted basement membrane in vitro. The layer occludes the pores of the membrane, blocking non-invasive cells from migrating through the membrane. In contrast, invasive cells (malignant and non-malignant) are able to detach themselves from and invade through the Matrigel Matrix.

Cells were seeded on Transwell Matrigel Chambers with reduced growth factors (BD Biosciences, Heidelberg, Germany) according to the manufacturer's protocol (25.000 cells) in serum free media. Afterwards, the cells are seeded on the top of the chambers on the matrigel layer and a chemoattractant is put into the bottom of the 24 well plates (in this case medium with serum was used). The plate is then incubated in a in a humidified tissue culture incubator, at 37oC, 5% CO_2 atmosphere for 6-22 hours depending on the experimental conditions (i.e. if measuring invasion or adhesion of cells). The cells migrate through the matrigel layer depending on their migrative potential. Non migrating cells are then removed and the remaining cells are stained (using crystal violet, Sigma) and counted under a standard light microscope. Data is expressed as the percent invasion through the Matrigel Matrix and relative to the migration through a control chamber (this is a non coated insert). For adhesion experiments, the same procedure was used with collagen coated chambers instead of matrigel chambers.

2.16 Colony Formation

Cells were seeded on 60mm dishes or 6-well cluster plates and grown for up to a week (1000-5000 cells per well) and fixed with 4% paraformaldehyde, stained with crystal violet, photographed and counted under the microscope at 20x magnification. To simulate hydrophobic surfaces, ultra low attachment plates and 60mm petri dishes were also used for colony formation experiments.

2.17 Statistics

Receiver Operating Characteristic analysis was performed in order to determine the optimal methylation (PMR value) cutoff. As result, PMR values above 30% were considered as methylation as methylated, whereas PMR levels below 30% were classified as unmethylated. This was counted

Methods

as PMR over 30% in tumor tissue and PMR below 30% in the adjacent non-neoplastic tissue for the initial cohort and PMR over 30% in tumor tissue for the other 4 patient cohorts. Correlations of the methylation event with clinicopathological features, such as primary tumor site, histological grade of differentiation or stage of cancer were assessed by the Fisher's exact test. The Mann-Whitney U test was used to compare the distribution of quantitative data between two independent samples.

To consider potential cluster effects by different study centers involved, random effect models were employed for statistical analysis of response probability in relation to PMR. Based on these models, estimates of response ratios (relative risks) were provided with 95% (fixed effect) confidence intervals. All tests were two-sided, and a p-value of <0.05 was considered statistically significant. Statistical calculations were done with Graphpad Instat 3, Graphpad Prism 5 (GraphPad Software, Inc., La Jolla, USA), MIX 1.7[175] software (see also[176]) and the R statistical software v2.9 (R Foundation for Statistical Computing, Vienna, Austria).

3. Goal and Purpose of this Thesis

As mentioned before (see introduction section), colorectal cancer is the second most common cancer in women and men in Germany and though colonoscopy is used for screening of benign and (pre)-malignant lesions (i.e. polyps) on persons over 55 years of age and paid for, compliance is low. Thus, molecular biomarkers could be useful for non-invasive diagnostic tests (i.e. blood and stool tests) to improve early detection rates but also for other clinical applications like prognostic factors, response prediction and follow-up monitoring, i.e. the likelihood of recurrence.[139] Epigenetic markers such as hypermethylated promoter regions of tumor suppressor genes, offer certain advantages over other molecular markers (DNA mutations, RNA and protein expressions levels) like high stability even in body fluids and detection by relative easy and robust methods (e.g. PCR based ones).[177]

Thus, the goal of this thesis was the evaluation and functional characterisation of some candidates from a panel of DNA methylation markers as useful biomarkers for clinical relevant applications including diagnosis, prognosis and response prediction in colorectal cancer. For this purpose, from a total of 12 markers originally discovered in a study[166] for the development of a diagnostic screening test in blood plasma, a couple were selected according to certain criteria (see results section below). As these genes are hypermethylated in tumor samples and thus probably transcriptionally silenced, they could represent potential tumor suppressor genes. Therefore the hypermethylation frequency of the selected markers was first screened in a cohort of patient tumor samples. The markers were then tested *in vitro* in human colon adenocarcinoma cell lines (which have no endogenous expression of these markers) for its function in the context of colorectal adenocarcinoma through forced re-expression by transfection of a plasmid carrying the coding sequence of a marker. Cell culture based assays were performed to see changes in cell proliferation, apoptosis, migration and adhesion, stress response as well as response to chemotherapeutic agents on a cellular level. Potential downstream targets and interaction partners of the marker genes were identified as well. The results were then correlated with the results from the patient screening and clinical characteristics of the tumors. Any clinical relevant findings were validated in additional patient cohorts, if possible.

4. Results

4.1 Markers from Epigenomics

In July 2007, we received a list of 11 marker candidates from Epigenomics AG (Berlin, Germany and Seattle, USA) which were mostly obtained from a Differential Methylation Hybridization[178, 179] microarray or by methylation-specific arbitrarily primed PCR and methylated CpG island amplification.[165] Differential Methylation Hybridization (DMH) is a high-throughput DNA methylation screening tool that utilizes methylation-sensitive restriction enzymes to profile methylated fragments by hybridizing them to a CpG island microarray. According to this procedure, the sample DNA is first digested using MseI or another enzyme that does not cut at CG-rich regions to reduce genome complexity (this can also be done using sonication). Then, an universal linker adapter is ligated to the ends of the DNA fragments and the mixture is digest with a methylation sensitive restriction enzyme (which only cuts on a CpG site if the cytosine is unmethylated, 5mC blocks the digestion) like BstuI or HpaII (or a mixture of two enzymes to decrease the likelihood of incomplete digestion by any one enzyme and to reduce the possibility of false positive results). The purified restricted fragments will serve as templates for a final linker-mediated PCR with a universal primer. In the next step, the amplified fragments are fragmented, labeled with fluorescent dyes (usually Cy 5 and Cy 3) and hybridized to microarrays. The limitations of this method are the dependence on regions that have a restriction site for the used enzymes and problems with cross hybridisation and normalization. This can be addressed by optimization strategies for both fragmentation and restriction to increase the methylation content information (and reduce genomic complexity) as well as using probe sets targeting fragments devoid of methylation sensitive restriction sites and methylation calibrators for normalization.

Likewise, methylation-specific arbitrarily primed PCR (AP-PCR) depends on the use of methylation sensitive restriction enzymes to identify differentially methylated regions of genomic DNA. The principle of this method is to design primers which flank a restriction site of such an enzyme (e.g. HpaII or MspI) and to then cut the DNA templates before amplification. Thus, amplification will only be possible if the DNA is methylated at the target site. As a control, a methylation insensitive enzyme (such as RsaI) can be used. The problems with this technique lie in the nature of the used enzymes (e.g. incomplete digestion) which affect sensitivity. However, this method can be combined with universal primers and sequencing for genome wide identification of

differentially methylated sequences in a high throughput manner. Methylated CpG island amplification (MCA) also uses restriction enzymes that have differential sensitivity to 5-methylcytosine and followed by adaptor ligation and PCR amplification; methylated CpG rich sequences are preferentially amplified. For MCA, unmethylated SmaI sites are eliminated by digestion with SmaI (which is methylation sensitive and cuts blunt end) and methylated SmaI sites are then digested with the nonmethylation-sensitive SmaI isoschizomer XmaI, which digests methylated cg sites and leaves sticky ends. Adaptors are ligated to these sticky ends, and PCR is performed to amplify the methylated sequences. The MCA amplicons can be used directly (e.g. analysed on an electrophoresis gel) or products can be used to clone differentially methylated sequences. All three methods are useful for genome wide discovery processes to identify sequences that are differentially methylated (and are based on PCR in principle). In this case, the comparison was between colorectal neoplasia, normal colon tissue, and peripheral blood lymphocytes (PBL) from healthy age-matched individuals as well as tissue from other cancers. The marker selection was based on using the following scoring variables: (*a*) appearance using multiple discovery methods; (*b*) appearance in multiple pools of like samples; (*c*) located within a CpG island; (*d*) located within the promoter region of a gene; (*e*) located near or within predicted or known genes; (*f*) known to be associated with disease; (*g*) class of gene (transcription factor, growth factor, tumor suppressor, oncogene); and (*h*) repetitive element. For validation of the identified markers, MethyLight assays were used (see Methods section of this thesis). However, some of the most promising markers (including 2 of the 3 studied in this thesis), were first selected by literature research or bioinformatic analysis (i.e. the number of CpG islands). For all markers, though, with the exception of TUSC3 gene, performance data was obtained from Epigenomics. This was done by DMH microarray cluster analysis and subsequent MethyLight assays on tissue from colorectal cancer patients and methylation levels were later compared to breast cancer, liver and prostate cancer cell lines and normal colon, breast, liver and prostate samples. The primer and probe sequences for all markers were provided by Epigenomics. The methylation markers were all protein coding genes in Homo sapiens, but the actual candidate sequences – i.e. the location of the analyzed amplicons were mostly in noncoding DNA, depending on the CpG island location (e.g., promoter, introns or untranslated regions of the first exon). With TUSC3, which was earlier selected due to previous work done together with Epigenomics by M.P.A. Ebert (group leader and principle inverstigator), this list had 12 potential marker genes.

Table 5 The following chart gives an overview for all candidate marker genes.

ENTREZ Gene ID	Official Symbol	Official Full Name (HGNC)	Source	CpG Islands	Exon count	Variants/ Isoforms	Amplicon Location
1910	EDNRB	endothelin receptor type B	DMH	1	7	3,2	Exon 1
2737	GLI3	GLI family zinc finger 3	DMH	2 (3)	15	1	Intron 1
375612	LHFPL3	lipoma HMGIC fusion partner-like 3	DMH	1	3	1	distal promoter
3984	LIMK1	LIM domain kinase 1	APPCR	1	15	1	Intron 2
57575	PCDH10	protocadherin 10	DMH	1	1,5	2	Exon 1
128674	PROKR2	prokineticin receptor 2	DMH	1 (2)	2	1	proximal promoter
5098	PCDHGC3	protocadherin gamma subfamily C, 3	DMH	2 (4)	1,4	3	Exon 1
9770	RASSF2	Ras association domain family member 2	Literature	1	11,12	2,1	proximal promoter
51289	RXFP3	relaxin/insulin-like family peptide receptor 3	DMH	1	1	1	proximal promoter
2040	STOM	stomatin	DMH	1	3,7	2	Exon 1
339488	TFAP2E	transcription factor AP-2 epsilon	Literature	2	7	1	Intron 3
7991	TUSC3	tumor suppressor candidate 3	Literature	1	10,11	2	Exon 1

4.2 Marker selection

The markers were first screened for mRNA expression (and in selected cases methylation) in a classical set of colorectal cancer cell lines to find markers where a suitable *in vitro* model existed. The cell lines that were used were CACO2, DLD1, HCT116, HT29, LOVO, SW480 and its metastasis derived counterpart SW620. HEK cells were used as a control, since they should express all of these genes. The rationale was that marker genes with methylated promoter or enhancer regions in samples of colorectal cancer patients are transcriptionally silenced (through epigenetic mechanisms with DNA hypermethylation being the mark for it) and therefore represent potential tumor suppressor genes. Human colon adenocarcinoma cell lines which have these genes methylated as well might thus serve as a model for functional analyses of these genes. Literature research and functional classification of the genes was also used for decision making. In this step, markers were selected by the following criteria: number of publications, known functions, published data for cancer, published methylation data, functional classification type. In the end, the three most interesting and promising markers were picked, because they showed a) no or weak expression in more than one cell line and strong reexpression after treatment with 5-azacytidine, giving a strong indication for hypermethylation; b) functional not well characterized, although

Results

knowledge about the basic function makes designing of functional experiments easier; c) not well described in the literature thus allowing novel findings and making publication of these findings easier; d) not linked to cancer or at least not linked to colorectal cancer thus being a new epigenetic marker in this field.

Table 6 Literature/Functional Classification Overview (as at December 2010).

Gene	Known Function(s)	Conserved Domains	Literature (no. of papers)	Marker for Cancer(s)	Lit. about Function (no. of papers)	Involved in Pathways	DNA Methylation Data
EDNRB	G protein-coupled receptor, mutations lead to Hirschsprung disease type 2	transmembrane receptor (rhodopsin family); Serpentine type 7TM GPCR chemoreceptor	125	neuroblastic tumors, esophageal squamous cell carcinoma, lung cancer, melanoma, leukemia, oligodendrogliomas, breast cancer, nasopharyngeal carcinoma, vulvar cancer	79	Melanogenesis, Neuroactive ligand-receptor interaction, Calcium signaling pathway	yes
GLI3	transcription factor, mediates Sonic hedgehog (Shh) signaling, activcates the patched Drosophila homolog (PTCH) gene	Zinc finger, C2H2 type SFP1; Putative transcriptional repressor regulating G2/M transition	73	Basal cell carcinoma, bladder cancer, prostate cancer, pancreatic cancer, breast cancer, colon cancer, multiple myeloma, B-cell lymphoma, Merkel cell carcinoma, endometrial carcinoma, heptacellular carcinoma, leukemia, blastomas, gastric cancer, basal cell carcinoma, skin tumors	32	Hedgehog signaling pathway, Basal cell carcinoma,	no
LHFPL3	Mutations result in deafness in humans and mice	Lipoma HMGIC fusion partner-like	8	Uterine leiomyoma	3	-	no
LIMK1	LIM motif mediates protein-protein interactions, and may be involved in brain development.	PDZ domain; LIM domain; Tyrosine kinase,catalytic domain;Protein Kinases,catalytic domain	64	Prostate cancer, breast cancer, cervical cancer, pancreatic cancer, gastric cancer	34	Regulation of actin cytoskeleton, Axon guidance, Fc gamma R-mediated phagocytosis	yes
PCDH10	cadherin-related neuronal receptor, plays a role in the establishment and function of specific cell-cell connections in the brain	Cadherin repeat domain	17	cervical cancer, colorectal, gastric, pancreatic cancer	8	-	yes
PROKR2	G protein-coupled receptor for prokineticins, promotes angiogenesis and induces strong gastrointestinal smooth muscle contraction	transmembrane receptor (rhodopsin family)	18	postate cancer, ovarian cancer, hepatocellular carcinoma	14	Signaling by GPCR	no
PCDHGC3	neural cadherin-like cell adhesion protein, plays a role in the establishment and function of specific cell-cell connections in the brain	two Cadherin repeat domains	13	none	1	-	no
RASSF2	unknown, but tumorsuppressive	Ubiquitin-like domain of Rasfadin	29	colorectal cancer, breast and lung cancer, oral squamous cell carcinoma, gastric cancer, hepatocellular carcinoma, prostate cancer, endometrial cancer, thyroid tumors, leukemias	18	-	yes
RXFP3	binds the hormon relaxin 3	transmembrane receptor (rhodopsin family)	42	1 (endometrial carcinoma)	13	G-protein coupled receptor signaling	yes
STOM	binds to and alters the gating of acid-sensing ion channels	Band7 stomatin-like; membrane protease subunit	31	-	11	-	no
TFAP2E	UNKNOWN	AP-2 domain	4	prostate cancer	-	-	yes
TUSC3	required for cellular magnesium uptake, protein N-linked glycosylation via asparagine, association with adverse pregnancy outcomes	Protein Disulfide Isomerase (PDIa) family domain, redox active TRX domain; OST3 / OST6 family domain	14	prostate cancer, larynx and pharynx carcinomas, ovarian carcinoma	5	Protein processing in endoplasmic reticulum, N-Glycan biosynthesis,	yes

Results

From the pure literature view, the most promising markers were LHFPL3, PCDHGC3, RXFP3, TFAP2E and TUSC3 because none of these genes are well characterized and not (yet) linked to colorectal cancer in any way. However, of these genes, only RXFP3, TUSC3 and TFAP2E showed weak or absent expression and strong reexpression after treatment with Azacytidine in at least 3 out of 6 screened cell lines. Screening of these panel of cell lines for methylation before and after treatment also revealed a very strong promoter hypermethylation in most cell lines for all 3 candidate genes, which in most cases decreased after treatment around 10%-25%. See tables and figures below.

Table 7 The following list illustrates the results of the cell line screening (3 best markers).

Cell line	Gene	Methylation	Methylation AZA	Expression	Expression AZA
HCT-116	TUSC3	Over 50%	Over 50%	weak	strong
HT-29	TUSC3	Over 25%	Over 15%	weak	strong
SW480	TUSC3	Over 25%	Over 15%	none	strong
CACO-2	TUSC3	100%	Over 75%	none	strong
DLD-1	TUSC3	100%	Over 75%	weak	strong
LOVO	TUSC3	100%	Over 75%	none	strong
HCT-116	TFAP2E	Over 75%	Over 75%	strong	strong
HT-29	TFAP2E	Over 75%	Over 50%	weak	strong
SW480	TFAP2E	100%	Over 25%	none	strong
CACO-2	TFAP2E	100%	Over 75%	none	strong
DLD-1	TFAP2E	100%	100%	weak	strong
LOVO	TFAP2E	100%	100%	weak	strong
HCT-116	RXFP3	100%	100%	strong	strong
HT-29	RXFP3	Over 75%	Over 50%	weak	strong
SW480	RXFP3	Over 25%	Over 15%	none	strong
CACO-2	RXFP3	100%	Over 50%	none	none
DLD-1	RXFP3	Over 75%	Over 75%	strong	strong
LOVO	RXFP3	100%	Over 15%	strong	strong

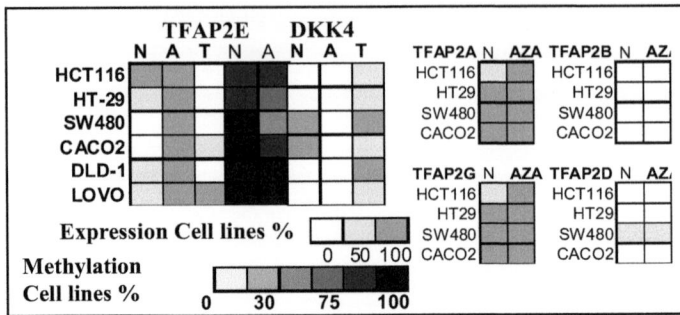

Figure 18 Expression and Methylation (heatmap related to percentage) of CRC cell lines untreated and treated with Azacytidine or Trichostatin A for TFAP2E and the downstream target gene DKK4 as well as the other AP-2 family transcription factors.

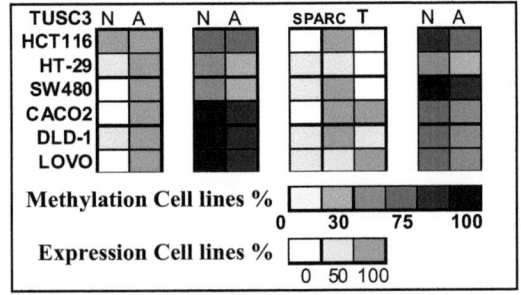

Figure 19 Expression and Methylation of CRC cell lines untreated or treated with Azacytidine (in the case of SPARC also treated with Trichostatin A) for TUSC3 and its downstream target gene SPARC.

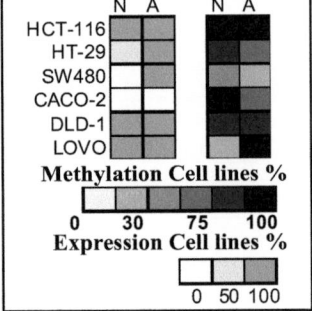

Figure 20 Expression and Methylation of CRC cell lines untreated or treated with Azacytidine for RXFP3.

Results

Table 8 Cell line screening continued (all other markers).

Cell line	Gene	Expression	Expression AZA	Gene	Expression	Expression AZA
HCT-116	PCDH10	strong	strong	RASSF2	strong	strong
HT-29	PCDH10	weak	strong	RASSF2	none	strong
SW480	PCDH10	strong	strong	RASSF2	strong	strong
CACO-2	PCDH10	none	none	RASSF2	strong	strong
DLD-1	PCDH10	strong	weak	RASSF2	weak	weak
LOVO	PCDH10	strong	weak	RASSF2	weak	weak
HCT-116	PCDHGC3	strong	strong	STOM	strong	strong
HT-29	PCDHGC3	strong	strong	STOM	none	strong
SW480	PCDHGC3	strong	strong	STOM	strong	strong
CACO-2	PCDHGC3	none	strong	STOM	none	strong
DLD-1	PCDHGC3	weak	strong	STOM	strong	strong
LOVO	PCDHGC3	strong	strong	STOM	none	strong
HCT-116	LIMK1	weak	strong	LHFPL3	strong	strong
HT-29	LIMK1	strong	strong	LHFPL3	weak	strong
SW480	LIMK1	weak	strong	LHFPL3	strong	strong
CACO-2	LIMK1	strong	strong	LHFPL3	none	none
DLD-1	LIMK1	strong	strong	LHFPL3	none	none
LOVO	LIMK1	strong	strong	LHFPL3	none	none
HCT-116	PROKR2	none	strong	GLI3	none	none
HT-29	PROKR2	none	none	GLI3	none	none
SW480	PROKR2	none	none	GLI3	strong	strong
CACO-2	PROKR2	none	none	GLI3	weak	weak
DLD-1	PROKR2	strong	strong	GLI3	none	none
LOVO	PROKR2	none	none	GLI3	strong	strong
HCT-116	EDNRB	none	strong			
HT-29	EDNRB	none	weak			
SW480	EDNRB	none	strong			
CACO-2	EDNRB	weak	strong			
DLD-1	EDNRB	weak	strong			
LOVO	EDNRB	none	strong			

4.3 Marker Evaluation

The selected markers were then screened for methylation levels in a cohort of 74 CRC patients from the University Hospitals Magdeburg and Berlin (148 samples, matched non-tumor and tumor tissue each, snap frozen material) with MethyLight assays. By using standard curves derived from artificially fully methylated human DNA and serial dilutions (1:4), methylation ratios were measured as percentage of this standard (PMR or percentage of methylation ratio). To determine the level of background methylation, Receiver Operated Characteristics Curves were applied to establish cutoff values for the markers which would best discriminate normal tissue from tumor based on the assumption that normal tissue has much lower methylation levels then tumor tissue.

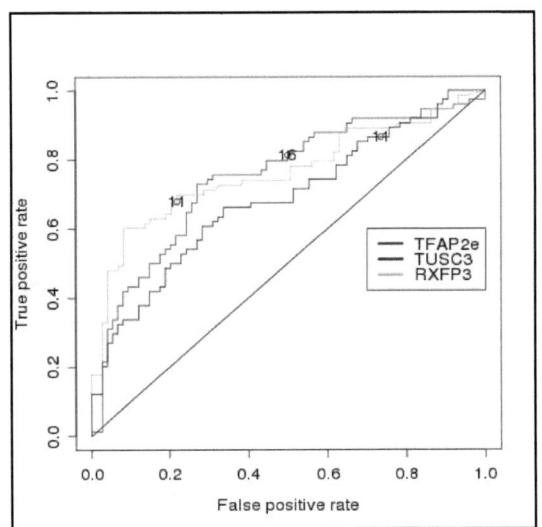

Figure 21 ROC Curves for all there selected markers. Sensitivity and specificity were maximized by choosing the cut point according to the maximal Youden index J.

Table 9 ROC results.

Marker	Cutoff as PMR value	AUC	Sensitivity (for matched tissue)	Patients classfied as postive	Patients classfied as negative	Total number of patients
TUSC3	52	0.68	0.35	26 (35%)	48	74
TFAP2E	30	0.78	0.51	38 (51%)	36	74
RXFP3	23	0.76	0.61	45 (61%)	29	74

With the established cutoff values, the patients were either classified as positive, i.e. these patients had lower methylation levels then the cutoff value in normal tissue and higher methylation levels in the tumor tissue or negative, if the methylation levels were higher or lower than the cutoff in both tumor and normal tissue. In rare cases, methylation levels were higher in normal tissue then in tumor tissue, though human handling errors could not be excluded (mixing up NT and TU samples), these samples were also classified as negative. If possible these samples were measured twice. For a detailed list of PMR values see the **Appendix A** of this thesis. The 148 samples were processed in a total of 18 DNA extractions (see Method section of this thesis). For 64 patients, clinical data regarding age, gender and TNM status were available. Therefore, the first 10 patients were numbered E1-E10 to denote them as extra samples. For TUSC3 and TFAP2E methylation levels differed drastically between normal and tumor tissue only in a subset of patients (about 35% for TUSC3 and 50% for TFAP2E). The majority of the patients classified as negative for TUSC3 and TFAP2E methylation showed a high methylation levels in normal as well as tumor tissue. For RXFP3 it could be said that at least more then half of the patients had high methylation levels in the tumor and low in the normal tissue, furthermore, most of the non-methylated patients showed almost no methylation in both tumor and normal tissue.

To determine if any of the three selected marker genes had a correlation with clinicial characteristics, the methylation results of the 64 patients for which the cliniocial data was available underwent a statistical analysis with the Graphpad Prism Program. The following characteristics were put into consideration: age, gender, tumor location (rectum, sigmoid, recto-sigmoid, colon, colo-sigmoid), TNM status, and grading as well as mucus production (i.e. tumor described as mucinous or not). No correlation with the measured clinical characteristics was observed in these samples (see figures below). Additionally, mutation analysis with a RanPlex CRC array was done to match methylation of TFAP2E, RXFP3 and TUSC3 with TP53, APC, BRAF and KRAS mutations in these tumors. For TFAP2E, mRNA expression was matched with those of its identified downstream target gene DKK4 in tumor tissue of 28 patients (see tables and figures below and **Appendix B**).

Table 10 Correlation of TFAP2E and DKK4 expression and methylation.

n = 28 (tumor)	DKK4 mRNA expressed	DKK4 mRNA not expressed	TFAP2E methylated	TFAP2E umethylated
TFAP2E expressed	6	2	5	3
TFAP2E not expresssed	11	9	12	8

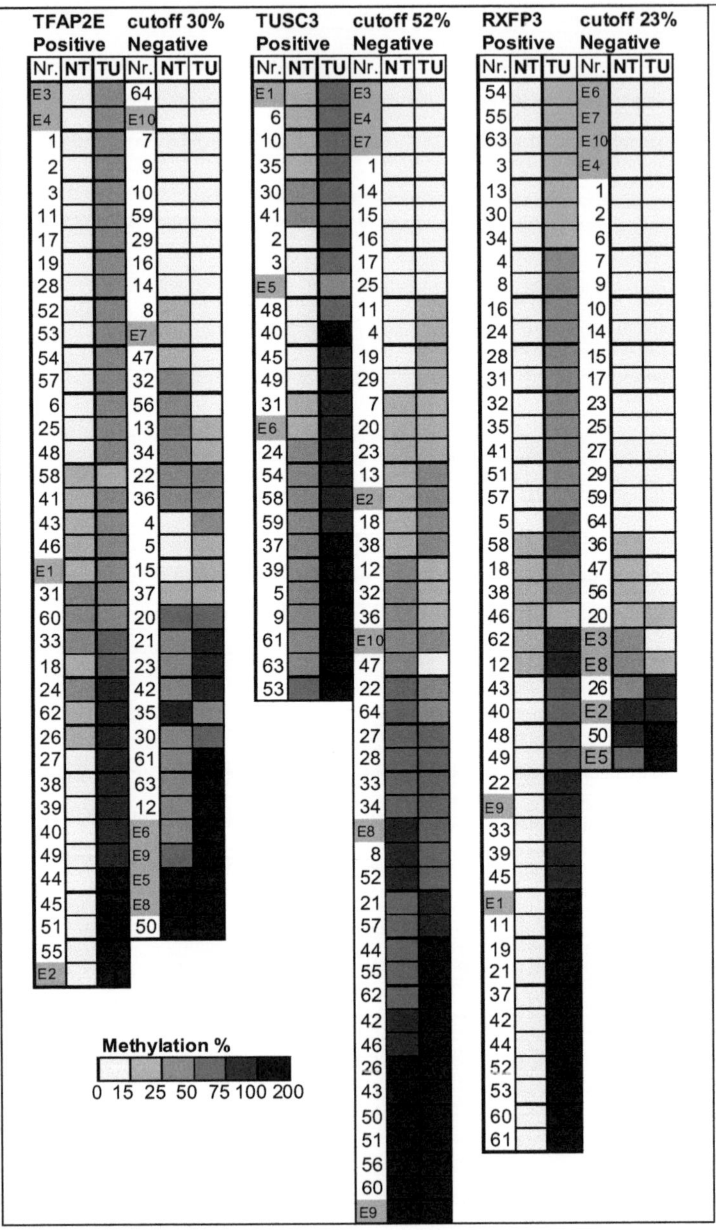

Figure 22 Methylation results of non-tumor (NT) and tumor (TU) tissue of 74 patients for the three selected markers TFAP2E, TUSC3 and RXFP3.

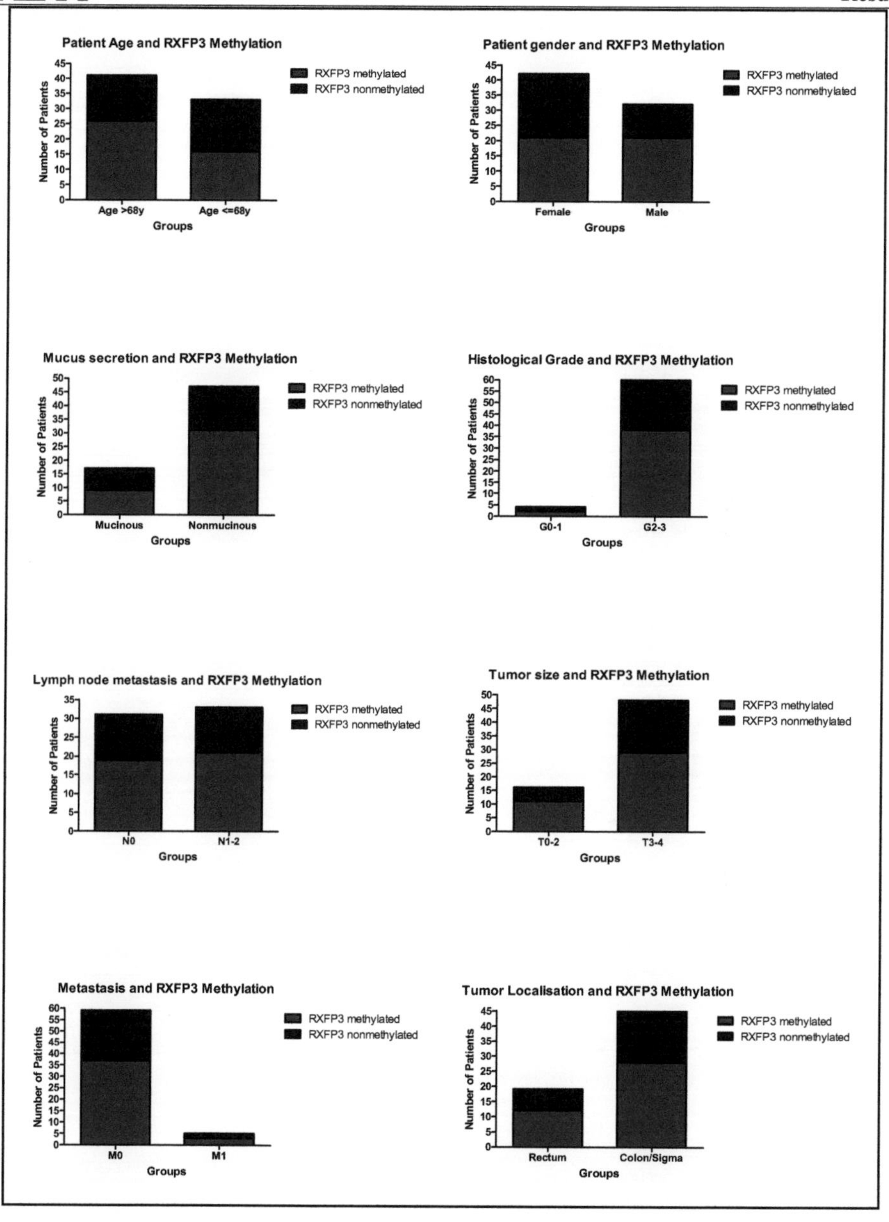

Figure 23 Statistical analyses for correlation with clinical characteristics for RXFP3 methylation in the 74 patients.

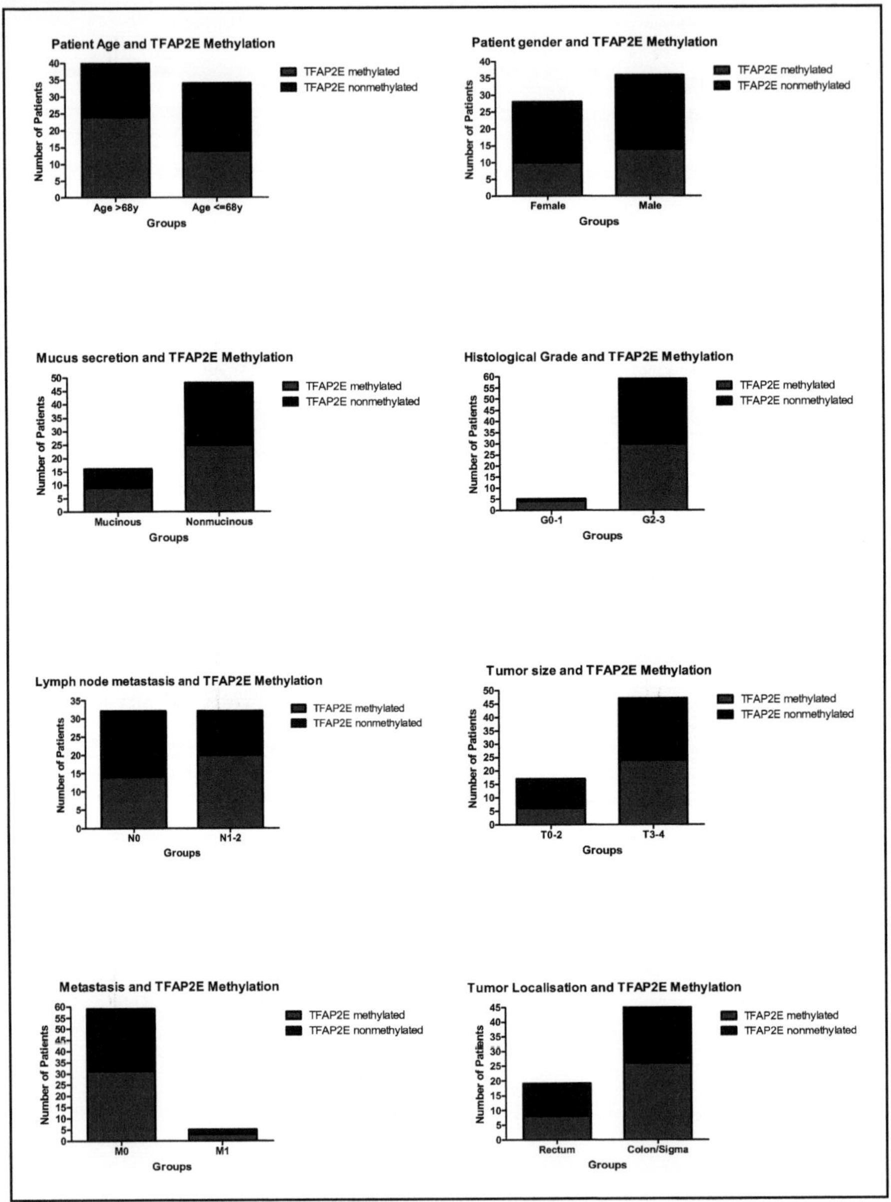

Figure 24 Statistical analyses for correlation with clinical characteristics for TFAP2E methylation in the 74 patients.

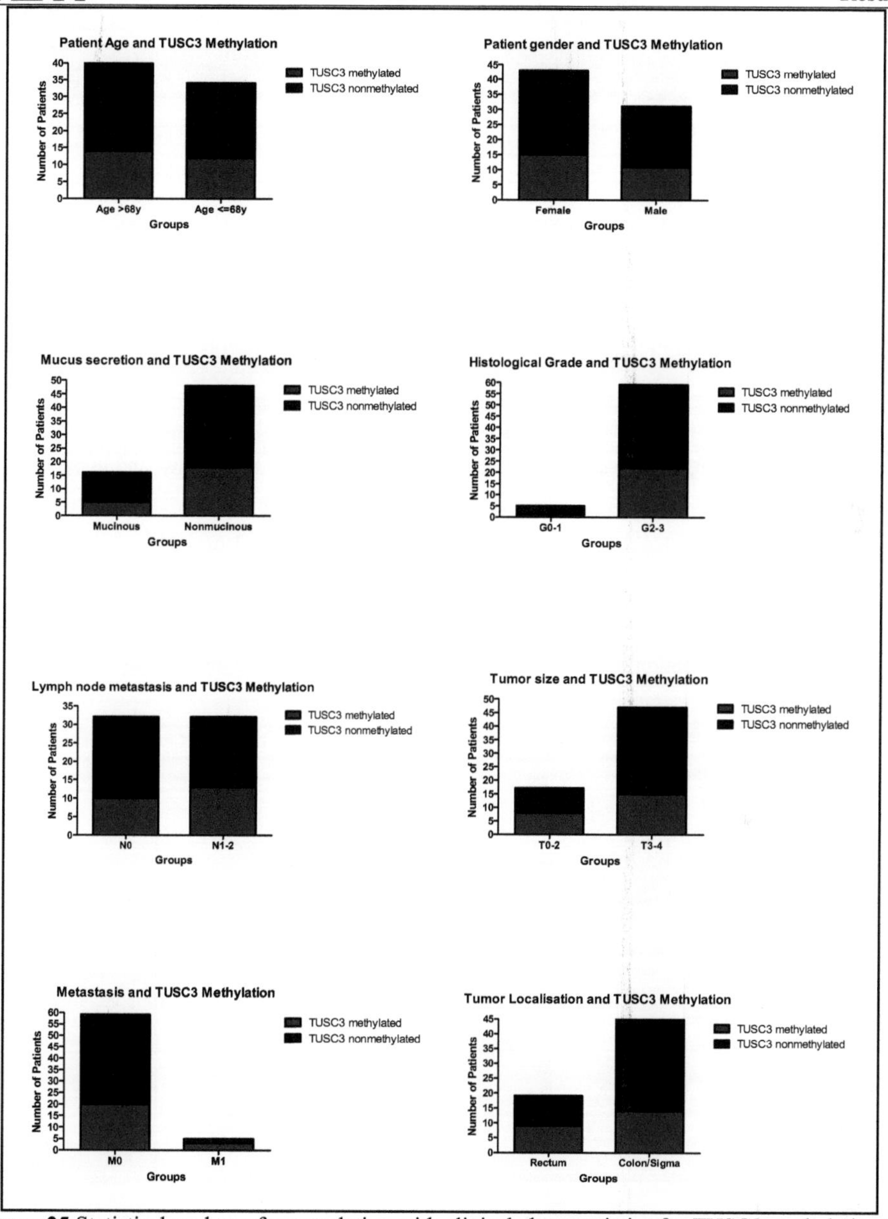

Figure 25 Statistical analyses for correlation with clinical characteristics for TUSC3 methylation in the 74 patients.

Table 11 Results of the mutation analysis for RXFP3, TUSC3 and TFAP2E.

Methylation Mutation	TUSC3 methylated	P value	RXFP3 methylated	P value	TFAP2E methylated	P value	Total
TP53	9	ns.	14	ns.	10	ns.	21
KRAS	3	ns.	9	ns.	8	ns.	11
APC	8	ns.	11	ns.	12	0.0077	16
BRAF	1	ns.	3	ns.	3	ns.	4
KRAS+APC	11	ns.	20	0.0012	20	0.0021	27
KRAS+TP53	12	ns.	23	ns.	18	ns.	32
APC+TP53	16	0.0068	25	ns.	22	ns.	37

Out of 63 patients (from 11 patients not enough material was obtainable for the Ranplex CRC arrays) for the APC, KRAS, TP53, BRAF genes and correlation with methylation of RXFP3, TUSC3 and TFAP2E respectively (see **Appendix B** for a detailed list of mutations for each patient sample). In 48 of the 63 patient samples, at least one mutation could be detected and in 50 samples also at least one marker gene was methylated (46 samples showed both methylation and mutation). A significant correlation between mutation and methylation was found for APC mutations versus TFAP2E methylation (p < 0.0077, Fisher's exact test, 99% CI) as well as APC plus KRAS mutations together (versus TFAP2E methylation, p < 0.0021 and versus RXFP3 methylation p < 0.0012) and the combination of APC plus TP53 mutations and TUSC3 methylation (p < 0.0068). See figure below.

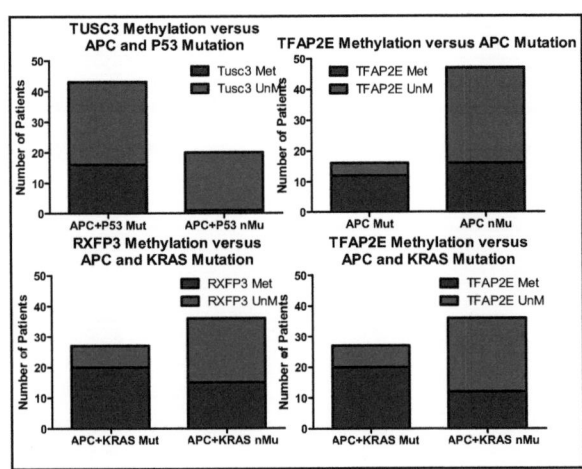

Figure 26 Significant correlations for mutations of APC and TP53 for the methylated marker genes TUSC3 and TFAP2E.

4.4 Functional Analysis

4.4.1 Cell clones

To analyze the effects of re-expression of the three selected markers in tumor cells, for each of the three genes SW480 cells were stable transfected with a pTarget plasmid containing the coding sequence of TUSC3, TFAP2E and RXFP3 respectively. Empty pTarget vectors were used as control for transfection and subsequent experiments. Single clones were picked and grown over several weeks (see appendix for an overview) in media with G418. A set of at least three clone lines and controls were used for experiments checking possible tumor suppressive functions related to apoptosis, proliferation and migration. To determine the optimal G418 dose for selection of positive cell clones, dose response experiments were done for the used cell lines (SW480 and LOVO) and a concentration of 500µg/ml was used (see table below).

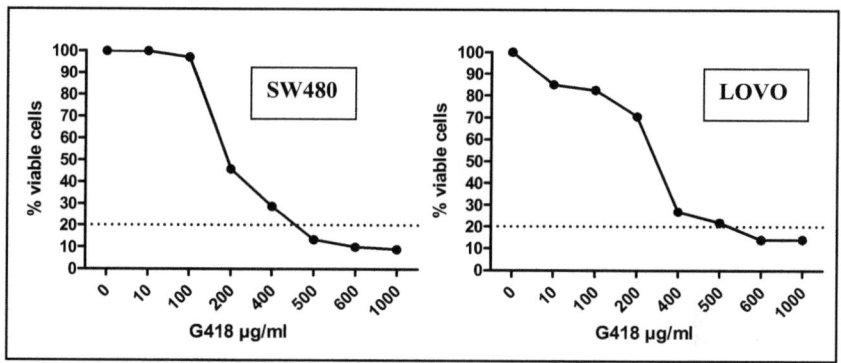

Figure 27 Dose response curves for determination of G418 concentration.

4.4.2 Microarray

To identify possible downstream interactions and target genes, Affymetrix Gene ST 1.0 Arrays were used to compare RNA expression levels of cell clones and empty vector controls for all three marker genes. All genes with a more than 3fold change in expression were checked for validation via quantitative PCR in at least 3 independent cells clones and 3 independent empty pTarget vector controls in every case. Additionally, from the 100 genes with the highest difference in expression (fold change) in the microarray(s), either the top 10 genes or potential interesting genes (via pathway analysis and literature search) were picked for validation via PCR and qPCR. The validation status of such potential downstream interaction partners of the three selected markers were classified as confirmed when all tested clones showed the same regulation as in the microarray (thus up or downregulation in all clones but not upregulation and downregulation in some clones). See tables and figure below and see also **Appendix A** for detailed lists of the first 100 up- and downregulated genes for each marker.

Table 12 Overview of downstream targets validation for TFAP2E overexpressing cells (Top 20 regulated genes).

Rank	Gene	Up/Down in Microarray	Confirmable via qPCR	Foldchange Microarray	Highest Foldchange Quantitative PCR
1	MAGEC2	UP	NO	23.97	0.95
2	ASB4	UP	NO	8.65	0.93
3	MUC13	UP	NO	3.12	19.6
4	SYTL3	UP	NO	2.82	no difference
5	SLC1A2	UP	NO	2.44	no difference
6	VIL1	UP	NO	2.43	4.99
7	GMFG	UP	NO	2.39	no difference
8	CKMT2	UP	NO	2.36	no difference
9	DDIT4	UP	NO	1.98	14.37
10	SERPINE1	UP	NO	1.97	14.2
1	**DKK4**	DOWN	**YES**	5.32	35.7
2	DKFZP564O0823	DOWN	NO	3.20	13.45
3	PSG5	DOWN	NO	2.52	no difference
4	ALDH1A3	DOWN	NO	2.48	0.75
5	PSG1	DOWN	NO	2.40	no difference
6	GABRA3	DOWN	NO	2.34	no difference
7	RBMY2EP	DOWN	NO	2.29	no difference
8	CSGALNACT1	DOWN	NO	2.24	no difference
9	PSG7	DOWN	NO	2.23	no difference
10	BNC2	DOWN	NO	2.15	no difference

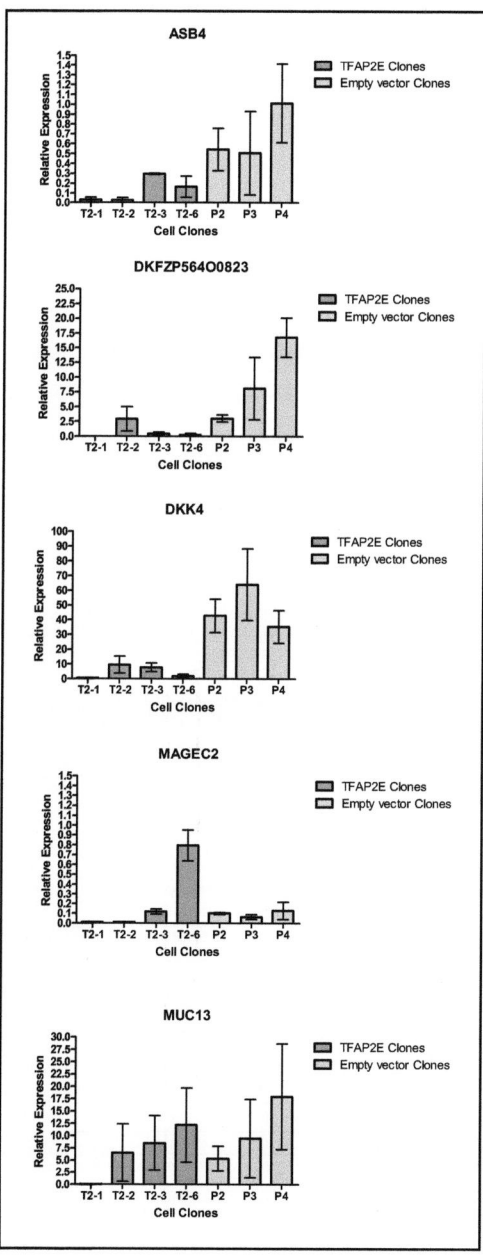

Figure 28 Validation of genes with a more than 3fold change in microarray via quantitative PCR.

Results

Table 13 Overview over the validation of downstream targets for TUSC3 overexpressing cells (Top 20 regulated genes).

Rank	Gene	Up/Down in Microarray	Confirmable via qPCR	Foldchange Microarray	Highest Foldchange Quantitative PCR
1	SPARC	UP	YES	10.88	3.78
2	HSPA6	UP	NO	5.74	no difference
3	ASB4	UP	NO	3.82	no difference
4	CD300A	UP	NO	3.28	no difference
5	CCR7	UP	NO	3.26	no difference
6	UIMC1	UP	NO	2.92	no difference
7	COL6A3	UP	YES	2.92	0.63
8	GEM	UP	NO	2.91	no difference
9	GPR35	UP	NO	2.77	no difference
10	NUPR1	UP	YES	2.72	1.38
1	PSG7	DOWN	NO	5.17	no difference
2	PSG5	DOWN	YES	5.03	0.72
3	KIAA1199	DOWN	YES	4.43	0.15
4	CYP24A1	DOWN	YES	4.11	0.22
5	PSG1	DOWN	NO	4.09	no difference
6	ITGB8	DOWN	NO	4.08	no difference
7	SEMA5A	DOWN	NO	4.07	no difference
8	TNC	DOWN	NO	3.87	no difference
9	SCNN1A	DOWN	NO	3.60	no difference
10	ANKRD36B	DOWN	NO	3.56	no difference

Table 14 Overview over the validation of downstream targets for RXFP3 overexpressing cells (Top 20 regulated genes).

Rank	Gene	Up/Down in Microarray	Confirmable via qPCR	Foldchange Microarray	Highest Foldchange Quantitative PCR
1	ROCK1	UP	NO	5.48	no difference
2	SCNN1A	UP	YES	4.32	5.45
3	LRRFIP1	UP	NO	4.21	no difference
4	DDIT4	UP	YES	4.09	14.15
5	PDGFA	UP	NO	4.01	no difference
6	DKK4	UP	NO	3.70	no difference
7	CA9	UP	YES	3.46	37.41
8	KCTD11	UP	NO	3.43	no difference
9	7A5 (MACC1)	UP	NO	3.35	no difference
10	ANKRD37	UP	NO	3.34	no difference
1	P2RY5	DOWN	NO	13.27	no difference
2	ZFAND2A	DOWN	YES	6.25	20.09
3	MAGEC2	DOWN	NO	4.85	no difference
4	DKK1	DOWN	YES	3.57	15.44
5	ARRDC4	DOWN	NO	3.40	no difference
6	SNAI2	DOWN	NO	3.27	no difference
7	PPP1R15A	DOWN	NO	3.16	no difference
8	DDIT3	DOWN	NO	3.04	no difference
9	EFCAB4B	DOWN	NO	2.96	no difference
10	TXNIP	DOWN	NO	2.78	no difference

Results

4.4.3 Cell Assays - TFPA2E

The cell clones were then investigated for changes in apoptosis and proliferation between overexpressing clones and empty pTarget controls.

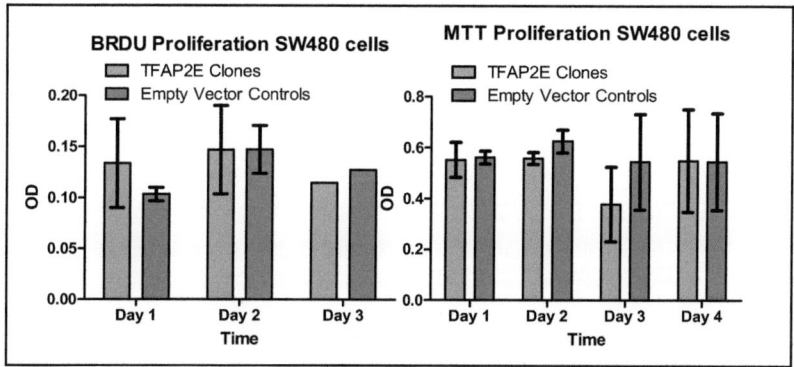

Figure 29 Proliferation of stable overexpressing TFAP2E SW480 cell clones measured via MTT and BRDU assays, mean of 3 independent experiments performed in triplicates, 5000 cells per well grown for 3-4 days.

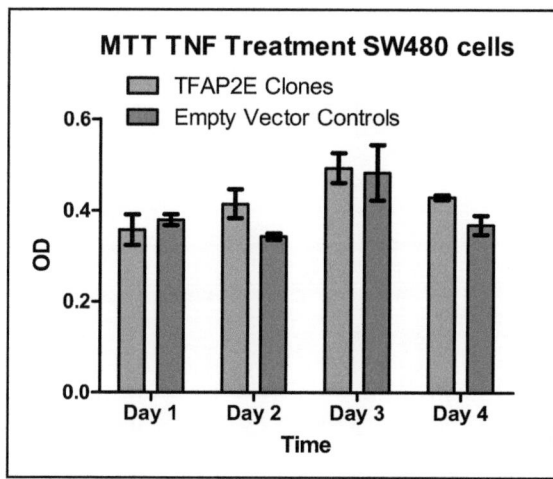

Figure 30 Apoptosis of stable Apoptosis overexpressing TFAP2E SW480 cell clones via TNF treatment and MTT assays, mean of 3 independent experiments performed in triplicates, 5000 cells per well grown for 3-4 days.

SW480 cells were also transiently transfected with pTFAP2E or pcDNA3-DKK4 and treated with oxaliplatin, irinotecan or fluorouracil (5-FU) for 3 days. Surviving cells were measured by MTT assays. All experiments were performed in triplicates. TFAP2E transfected SW480 cells showed a significant decrease (about 20% compared to controls) and DKK4 transfected SW480 cells showed a significant increase in survival (about 10% compared to controls) when treated with 5-FU after 2 days (Kruskal-Wallis test, $p < 0.01$). Cells transfected with both pTFAP2E and DKK4 exhibited intermediate responsiveness. In addition, these results were confirmed in SW480 cells clones stably transfected with TFAP2E or pTarget empty vector as control, which were treated with 5-FU (Mann-Whitney U test, $p < 0.001$). The optimal 5-FU concentration was determined by dose-response curves for TFAP2E and DKK4 transfected cells and using the approx. IC50 as working concentration. For irinotecan and oxaliplatin, no striking differences were observed at the initial used standard concentrations based on literature research.

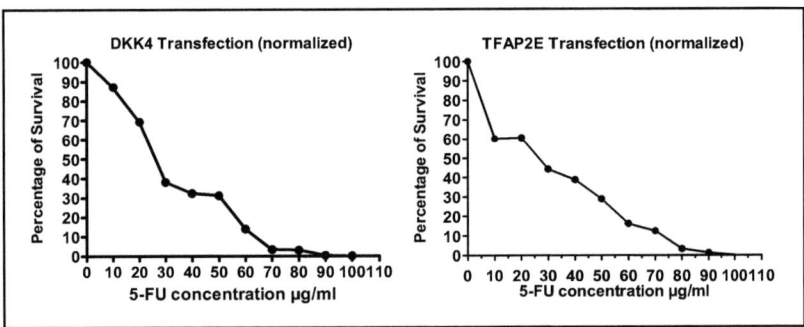

Figure 31 Dose response curves for SW480 cells to determine the optimal 5-FU concentration.

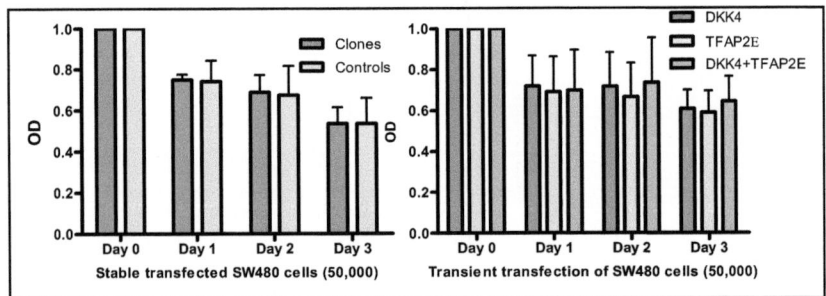

Figure 32 TFAP2E stable overexpressing clones or empty vector controls and SW480 cells transiently transfected with DKK4-CDS, TFAP2E-CDS or both treated with treated with 60 µM oxaliplatin.

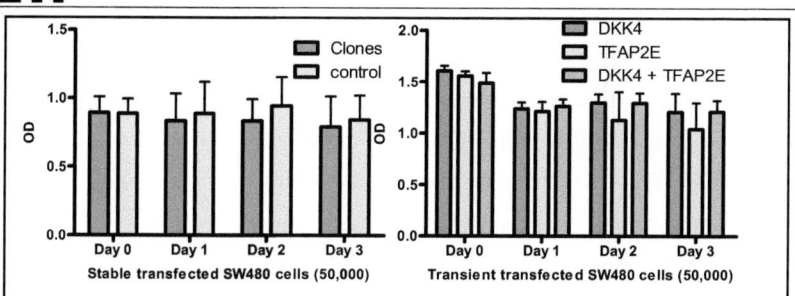

Figure 33 TFAP2E stable overexpressing clones or empty vector controls and SW480 cells transiently transfected with DKK4-CDS, TFAP2E-CDS or both treated with 20 µM irinotecan.

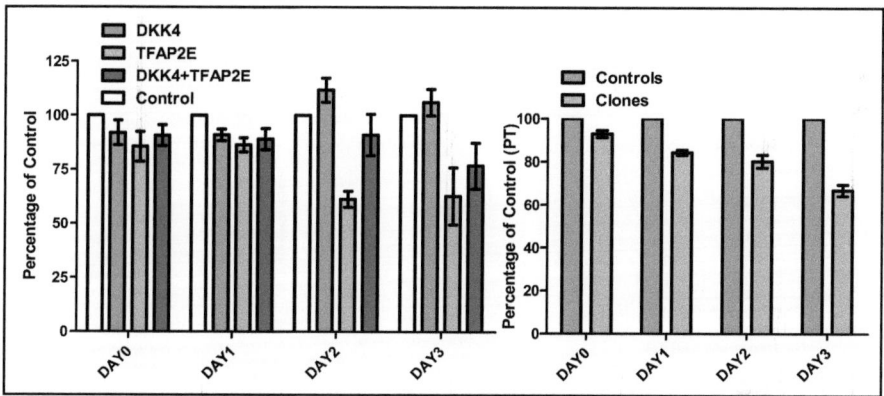

Figure 34 Resistance to treatment with fluorouracil (5-FU) irinotecan and oxaliplatin, mean of 3 independent experiments performed in triplicates (transiently transfected SW480 cells and TFAP2E stable overexpressing SW480 cell clones). SW480 cells transiently transfected with DKK4-CDS, TFAP2E-CDS or both (50,000 cells per well, treated daily with 50 µg/ml fluorouracil for a duration of 3 days) and TFAP2E stable overexpressing clones or empty vector controls 4 clones per group, 50,000 cells per well.

A search using the Transcription Element Search System (TESS) as a bioinformatic approach (http://www.cbil.upenn.edu/cgi-bin/tess) found 2 putative binding sites for AP-2 proteins within 2kb upstream and 2 sites within 1kb upstream of the DKK4 mRNA transcription start site. To analyze binding of TFAP2E to these sites, 1kb and 2kb of the DKK4 promoter were cloned into a Luciferase reporter vector (pGL3) and transfected together with pTFAP2E and pRL-TK (internal luciferase control) into SW480, CACO-2 (no TFAP2E expression) and HT-29 (weak TFAP2E expression) cells. Luciferase activity was decreased in CACO-2 cells transfected with TFAP2E compared to cells without TFAP2E transfection, 3-fold with the pGL3-DKK4-1kb vector and 5-fold with the pGL3-DKK4-2kb vector. SW480 cells showed the same effect with 3- to 7-fold downregulation (7-fold with pGL3-DKK4-2kb vector, Figure 1B) as well as in HT-29 cells (3-fold with pGL3-DKK4-1kb vector, Wilcoxon test $p < 0.005$). HEK cells showed no difference in luciferase activity with or without TFAP2E transfection. This might be due to the high endogenous TFAP2E expression level. See figure below.

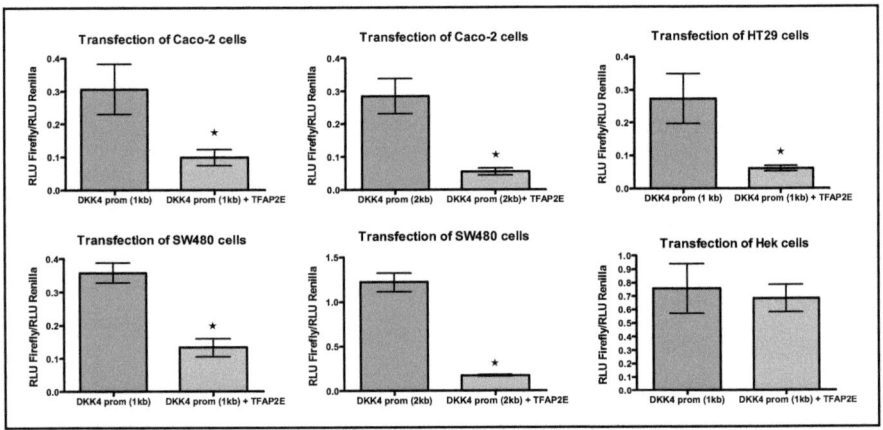

Figure 35 Luciferase reporter assays for CACO-2 and SW480 cells transfected with DKK4 promoter (1kb and 2kb fragments) and TFAP2E CDS as well as HT29 transfected with DKK4 promoter (1kb fragment) and TFAP2E CDS and HEK cells used as controls.*$p<0.01$.

To test for direct binding of the TFAP2E protein on the DKK4 promoter, chromatin immunoprecipitation was performed with SW480 cells stably transfected with TFAP2 and tagged with the Flag TAG. The recovered DNA of the input control, negative control and the SW480 clones was used for real time PCR with two primer pairs flanking 4 putative AP-2 protein binding sites in the DKK4 promoter (two sites 2kb upstream and two sites 1kb upstream of the DKK4 transcription start site), respectively. While the negative control showed no amplification, both clones showed a 4-fold higher amplification for the two sites 2kb upstream and 1.5-fold higher for the two sites 1kb upstream, indicating a direct binding of TFAP2E protein to the DKK4 promoter.

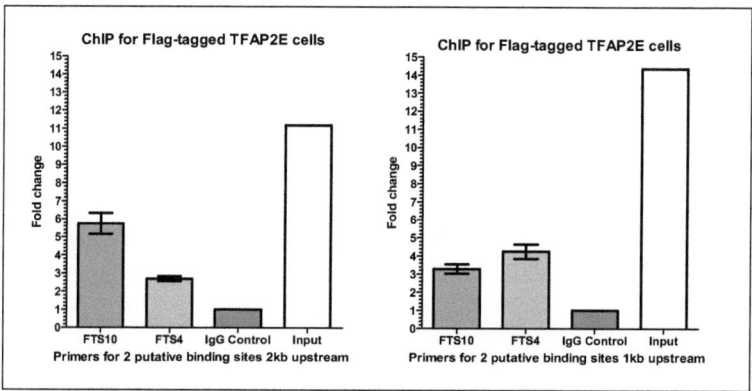

Figure 36 Chromatin immunoprecipitation relative quantification results for FLAG tagged stable TFAP2E overexpressing clones with two putative AP2 protein binding sites 1kb and 2kb upstream in the promoter of the DKK4 gene.

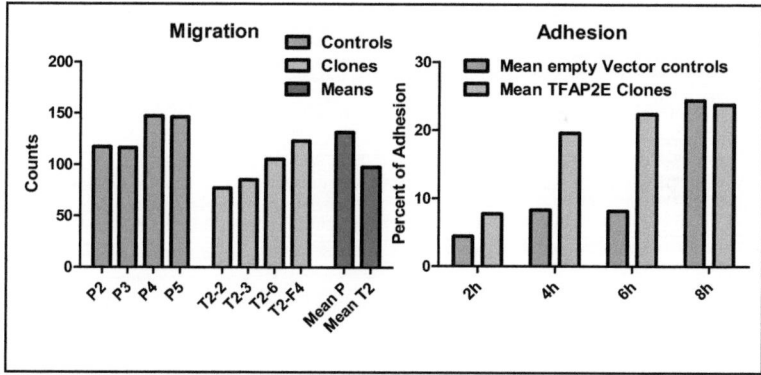

Figure 37 Migration and Adhesion Assays - Counts (absolute numbers) of migrating SW480 stable clones overexpressing TFAP2E and empty vector controls (25000 cells per well) through a matrigel chamber with reduced growth factors for 18h. Counts of adhering cells (percentage) of SW480 cell clones and empty vector controls (10000 cells) on noncoated plastic surfaces.

4.4.4 Cell Assays - TUSC3

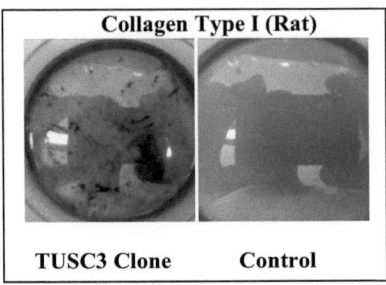

Figure 38 Adhesion of TUSC3 clones versus empty pTarget vector control clones on a collagen Type I coated plastic surface (example).

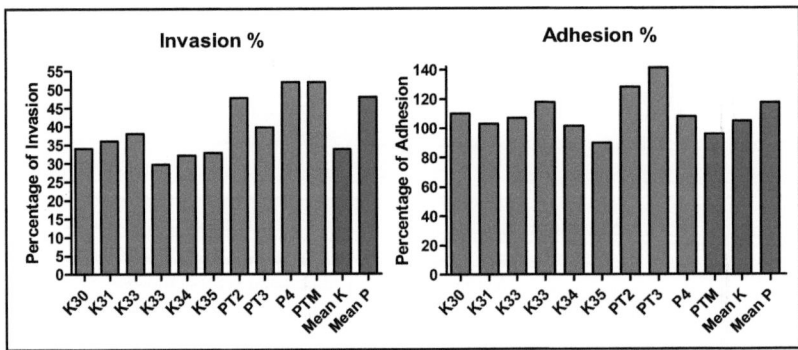

Figure 39 Percent of migrating cells through a matrigel chamber with reduced growth factor versus control chambers (invasion indices) for stable overexpressing TUSC3 SW480 cell clones (left side). Percent of adhering TUSC3 SW480 cell clones on a collagen coated (collagen Type I, human) matrix chamber versus control chambers (adhesion indices). Mean of two independent experiments with 5000 and 10000 cells per well for 18 hours (right side).

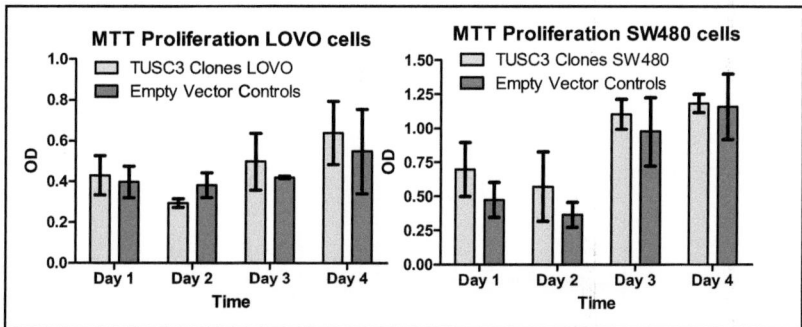

Figure 40 TUSC3 – Proliferation of stable overexpressing TUSC3 SW480 and LOVO cell clones measured via MTT assays, mean of 3 independent experiments performed in triplicates, 5000 cells per well grown for 3-4 days.

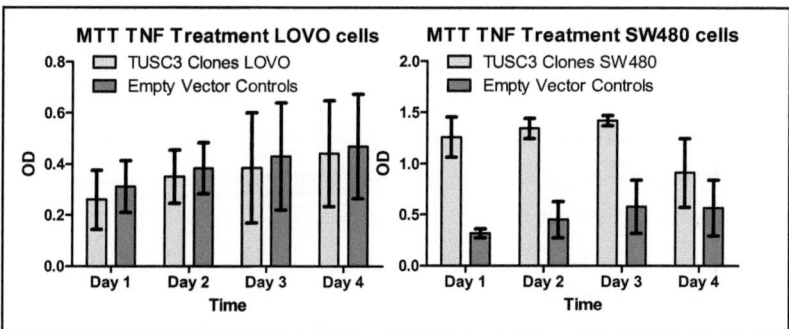

Figure 41 TUSC3 – Apoptosis of stable overexpressing TUSC3 SW480 and LOVO cell clones via TNF treatment and MTT assays, mean of 3 independent experiments performed in triplicates, 5000 cells per well grown for 3-4 days.

4.4.5 Cell Assays RXFP3

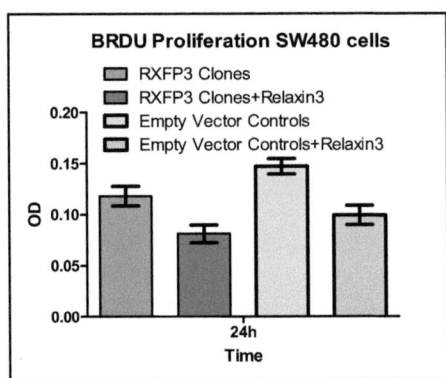

Figure 42 Proliferation of stable overexpressing RXFP3 SW480 cell clones measured via BRDU assays, mean of 3 independent experiments performed in triplicates, 5000 cells per well grown for 3-4 days.

4.5 Marker Validation (TFAP2E)

For validation of potential clinical use, the usefulness of TFAP2E methylation as potential biomarker for treatment response to fluorouracil was evaluated in another 4 independent patient cohorts (total number of patients = 220). The samples were from the University Hospitals in Munich (n=74), Mannheim (n=42), Bochum (n=74) and Dresden (n=36) mainly from patients enrolled in clinical trials there and consisted mostly of either paraffin embedded sections or DNA derived from these sections, except for 28/68 samples from Munich which where snap frozen biopsies. The cohorts were first analysed with MethyLight and then with High Resolution Melting analysis (HRM) with exception of the cohorts from Dresden and Bochum. In the case of Dresden only MethyLight analysis was performed since insufficient material was available for both. In the case of Bochum, only the first batch of the samples were analysed with both MethyLight and HRM, the rest were analysed with HRM only, since HRM proved to be more sensitive for FFPE samples. For the cohorts from Bochum, Mannheim and Munich, the results of the HRM analysis were used for validation, since the results from MethyLight assays proved to be unreliable in samples from formalin-fixed, paraffin-embedded tissue (see Appendix C) at least for the cohort from Mannheim. For cross-validation of the cutoff value, since it was applied to tumor tissue only, receiver operator

Results

characteristics curves were also generated for two of the cohorts were PMR values were available. Meta analysis were performed to estimate relative risks for each cohort and all together for responders and nonresponders alike and graphically represented via Forest Plots. See figure and table below.

Table 15 Validation results for TFAP2E in 4 independent patient cohorts.

Cohort No Center name No of samples	Analysis Method	Cancer Type	Response Evaluation	Methylation TFAP2E	Responder	Non-Responder	Fisher's exact test
I=Bochum n=74	HRM	Metastatic CRC	RECIST	Methylated	3	17	p < 0.0005
				Unmethylated	33	21	
II=Dresden n=36	MethyLight	Metastatic CRC	RECIST	Methylated	1	22	p < 0.0001
				Unmethylated	13	0	
III=Mannheim n=42	HRM	Primary Rectal Ca	Histology	Methylated	5	14	p < 0.0001
				Unmethylated	20	3	
IV=München n=68	MethyLight HRM	Primary Rectal Ca	Histology	Methylated	3	28	p < 0.0001
				Unmethylated	29	8	
I+II=RECIST n=110	HRM+ML	mCRC	RECIST	Methylated	4	39	p < 0.0001
				Unmethylated	46	21	
III+IV=Histology n=110	HRM	RC	Histology	Methylated	8	42	p < 0.0001
				Unmethylated	49	11	
I+II+III+IV Total (n=220)	HRM+ML	mCRC+RC	Both	Methylated	12	81	p < 0.0001
				Unmethylated	95	32	

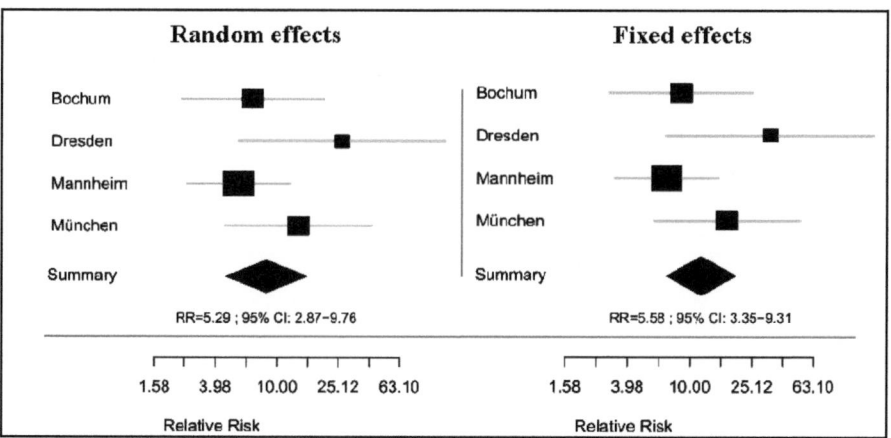

Figure 43 Meta-analyses on relative risks across all study centers were performed both by random and fixed effects models (95% confidence intervals). Both approaches report a significant overall effect with a relative risk factor of more than five (5.29 and 5.58 respectively) to respond to treatment when a patient shows - by definition - no TFAP2E methylation.

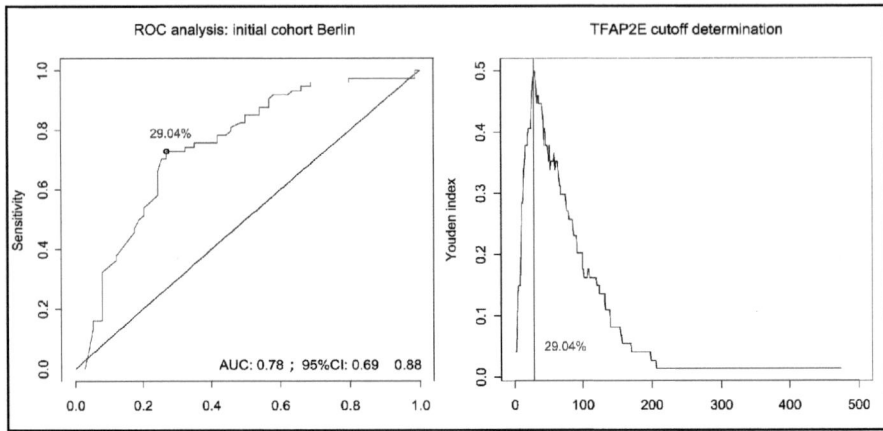

Figure 44 A patient-matched Receiver Operating Characteristics analysis was performed on the initial cohort from Berlin in order to define the optimal cutoff for TFAP2E methylation values (PMR). Sensitivity and specificity were maximized by choosing the cut point according to the maximal Youden value (right figure). The optimal cutoff was thus defined as 29.04% or PMR.

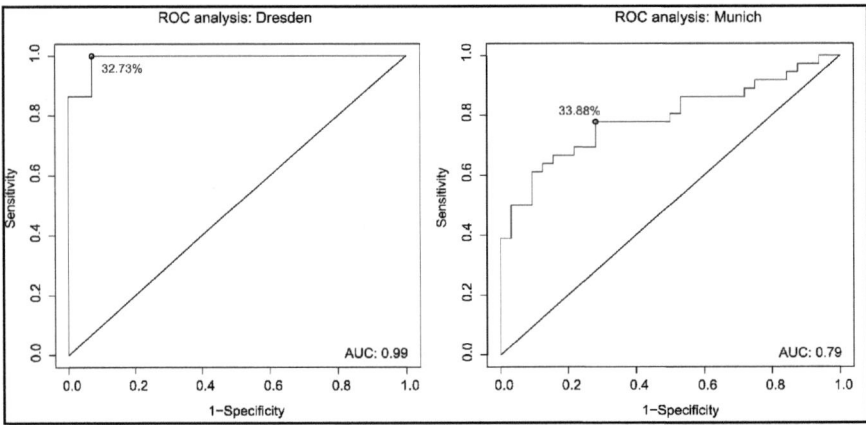

Figure 45 Receiver Operating Characteristics analyses were done on TFAP2E methylation values between responder and non-responder groups from Dresden and Munich. Optimal cutoffs were found for Dresden at 32.7% and for Munich at 33.9% (PMR).

5. Discussion

5.1 General

The goal of this thesis was the evaluation of a set of given DNA methylation markers as useful biomarkers for clinical relevant applications like diagnosis, prognosis and response prediction. From a panel of twelve, three promising markers (TUSC3, TFAP2E and RXFP3) were chosen according to literature as well as cell line data and analysed further in patients and *in vitro* studies. In the end, each marker could be associated with a different clinical setting. TUSC3 and TFAP2E performed poorly as diagnostic markers but this gave an indication that the markers define other features of colorectal adenocarcinomas. The high degree of methylation in normal tissue for TUSC3 and TFAP2E could indicate either an age related effect on methylation of these marker genes in colon mucosa (e.g. by epigenetic drift or random seeds of methylation) or an epigenetic field defect. RXFP3 showed no such methylation of normal tissue, therefore it can be assumed that methylation of RXFP3 is an event which occurs during tumorgenesis and RXFP3 could therefore be classified as a diagnostic factor for colorectal cancer. RXFP3 methylation was also associated with a combination of KRAS and APC mutations and most of the identified target genes which could be validated via qPCR were factors for hypoxia. TUSC3 methylation showed a correlation with APC and TP53 mutations as well as lymph node invasion and cell migration and adhesion. Thus, it could be useful as a prognostic factor since lymph node invasion is a key factor in metastasis formation. The identified downstream target of TUSC3, SPARC has also been linked to adhesion and migration as well as therapy resistance and patients with loss of SPARC expression show a similar overall survival as patients with loss of TUSC3 expression. The best characterized marker of this study, TFAP2E proved to be a tool for response prediction for patients treated with 5-FU.

5.1.1 Markers TFAP2E, TUSC3, RXFP3

Especially for advanced stages, resensitizing chemoresistant tumors to clinical approved drugs (like oxaliplatin and 5-Flourouracil) becomes an important goal.[180, 181] Novel candidate genes and mechanisms to predict chemoresponse for Oxaliplatin[182], Irinotecan[183] and 5-FU[184] are a major topic for advanced stages of colorectal cancer. Mutations in the enzymes which catabolite the drugs are mostly studied and DNA repair pathways, but can only explain some observed effects.[180, 185]

Methylation based markers (hypermethylated CpG-Island promoter regions of tumorsuppressor genes) seem to be most promising in both ways, since they would allow non-invasive feasible blood or stool tests. In recent years, some alternate genes which contribute to chemoresistance in colorectal cancer have been identified, including matrix metallopeptidases as MMP7[186], phosphoserine aminotransferases as PSAT1[187] and even the DNA polymerase POLB[188] for oxaliplatin and several genes for irinotecan[189, 190] including the epidermal growth factor receptor EGFR[191] and the ATP-binding cassette transporter gene ABCG2.[192] Genes involved in the TP53 mediated DNA repair pathway seem to play in important role[193-195] for oxaliplatin resistance as well. For 5-FU resistance, clinical features like CIMP and microsatellite instability are discussed to have a predictive value.[180, 196-199]

The TFAP2E gene itself lies on chromosome 1p34, covers 21.96kb, and spans 7 exons which encode a protein of 45.3 kDa.[200] Interestingly, this chromosomal region is deleted in several cancers, including colorectal cancer.[201] AP-2 proteins can bind to keratin promoters and act as hetero or homodimers. The AP2-transcription factor family consists of five members and plays important roles in developmental (namely eye, skin and neural structures) and cancer biology.[202] Expression of AP-2 proteins has been reported in various tissues, including brain and nerves, and notably breast for TFAP2C and skin for TFAP2E. TFAP2A is a known tumor suppressor in various cancers[203, 204] interacts with E-cadherin (CDH1)[205], the vascular endothelial growth factor VEGF[206] (a glycosylated mitogen that specifically acts on endothelial cells and has various effects, including mediating increased vascular permeability, inducing angiogenesis, vasculogenesis and endothelial cell growth, promoting cell migration, and inhibiting apoptosis which is often dysregulated in cancers and anti-VEGF therapies are important in the treatment of certain cancers and in age-related macular degeneration), APC[207] and TP53.[208, 209] TFAP2B is involved in several disorders[210, 211], mutations lead to Char-Syndrome[212] and seems to play a role in diabetes.[213, 214] While TFAP2C is induced by estrogen[215] and acts as a tumorsuppressor in breast cancer[203] by binding to ERalpha[216] and may predict tamoxifen resistance and being related to ERBB2[217, 218] (also known as Human Epidermal growth factor Receptor 2, Her2/neu, important as a drug target in breast cancer), not much is known so far about TFAP2D and TFAP2E. Complete knockout of TFAP2A, B and C is embryonic lethal in mice, knockout of TFAP2A leads to neural tube defects and other abnormalities.[219]

So far no transgenic mice exist for TFAP2D and TFAPE. All TFAP2 family members except TFAP2B have CpG Islands in their genomic structure, suggesting a role for epigenetic regulation of these genes. Indeed, TFAP2A has been shown to be hypermethylated in renal cell carcinoma[220] and

B-Cell lypmphoma[221] and similar results have been recently published for TFAP2C in breast cancer[222] and TFAP2E in prostate cancer.[167] In prostate and colorectal cancer, TFAP2D and TFAP2E have been identified as novel methylation markers[223] in global screening methods. However, no study has been done so far regarding the function of these markers. Since TFAP2D is not or just low expressed in colon[224], a major function in colorectal cancer seems unlikely compared to TFPAP2E. Here we report a first functional analysis linking TFAP2E to WNT-Signaling and chemoresistance in Colorectal Cancer Cells. TFAP2E has 2 CpG islands (one in the promoter region and on in intron 3), but the first one is methylated in blood lymphocytes suggesting tissue specific methylation patterns. In order to characterize the down-stream targets of TFAP2E we performed a microarray analysis of TFAP2E overexpressing cells and found DKK4 to be a potential target gene, which was significantly down regulated via TFAP2E. Further studies confirmed repression of DKK4 promoter activity through TFAP2E and binding of TFAP2E to the DKK4 promoter *in vitro*. DKK4 is a member of the dickkopf family, comprising various antagonists of WNT-signaling by binding to WNT-coreceptors LRP5/6.[225] DKK4 overexpression has been observed in the colon mucosa of patients with colitis.[226] However, the precise role of DKK4 in the colon mucosa and its contribution to carcinogenesis is so far unknown. The observation of a significant correlation of APC and KRAS mutations with TFAP2E methylation in cohort of 63 patients from this thesis (see results section) points to a connection between AP2 transcription factors and WNT signaling. For therapeutic uses, 5-Fluoruracil is the main component of polychemotherapies in colorectal cancer. Through addition of further drugs, including irinotecan and oxaliplatin, substantial improvement of progression-free and overall survival of patients with colorectal cancer has been achieved. Recently, targeted drugs, targeting the EGFR and VEGF, have also been approved for the first-line treatment of patients with colorectal cancer. Since overall survival of patients with CRC has increased from a median of 6-8 months in the era of 5-FU monotherapy up to 24 months in patients receiving polychemotherapies with targeted drugs, administration of these drugs in several lines of treatment is becoming a common strategy. Thus, individualisation of cancer therapy is becoming routine and drugs need to be chosen on an individual basis with respect to certain clinical and genetic response predictors. Accordingly, recent studies have demonstrated that patients with mutant K-ras cancer cells do not respond to EGFR inhibition. However, further molecular mechanisms underlying chemoresistance or allowing response prediction in this and other cancers are largely unknown.

5-FU is a thymidylate synthase inhibitor and therefore blocks synthesis of thymidine, leading to inhibition of DNA replication and inducing cell cycle arrest and apoptosis. 5-FU acts also as a pyrimidine analogue, is incorporated into DNA and RNA, inhibits replication enzymes and is transformed inside the cell into different cytotoxic metabolites. The main enzymes which mediate these anti-tumor effects are thymidine phosphorylase (TYMP) for metabolising the prodrug capecitabine and thymidylate synthetase (TYMS) for blocking of DNA synthesis; uridine-cytidine kinase 2 (UCK2), uridine monophosphate synthetase (UMPS or OPRT) and uridine phosphorylase 1 (UPP1) for RNA disruption; and methylenetetrahydrofolate reductase (MTHFR) for incorporation into DNA. Dihydropyrimidine dehydrogenase (DPYD) also plays a role in the thymidine and uracil pathway.[180] Resistance to 5-FU can, thus, be partly explained by mutations or polymorphisms in respective genes encoding these drug catalyzing enzymes, including thymidylate synthetase (TYMS), dihydropyrimidine dehydrogenase (DPYD) and methylenetetrahydrofolate reductase (MTHFR).[180] For example, polymorphisms or mutations in TYMS lead to increased 5-FU resistance in colorectal cancer and mutations in DPYD[227] and polymorphisms in MTHFR[228] lead to an increased risk of toxicity in cancer patients receiving 5-fluorouracil chemotherapy. Other genes that have been implicated in the mediation of 5-FU response include heat shock proteins as HSP27[229], HSP70[230], the hypoxia-inducible factor HIF1A[231] (a transcription factor which plays an essential role in cellular and systemic homeostatic responses to hypoxia), the cytoplasmic protein tyrosine kinase PTK2[232] (also known as focal adhesion kinase FAK), the zinc finger gene ZKSCAN3[233] and the apoptosis inhibitor BIRC3[234] as well as glycolytic enzymes like PKM2[235] and antioxidant enzymes like NFE2L2[236] have been identified. Interestingly, epigenetic factors such as HDACs, MBD4[237] and histones itself[238] seem to play a role in 5-FU resistance as well[239] by regulating or interacting with thymidylate synthase expression. In addition to DKK4, the genes SMYD3, SERTAD1, RRM2 (a reductase which catalyzes the formation of deoxyribonucleotides from ribonucleotides), ORC6 (part of a highly conserved six subunit protein complex essential for the initiation of the DNA replication) and EIF4E (a eukaryotic translation initiation factor involved in directing ribosomes to the cap structure of mRNAs) were found in the same study[237]. In addition, several methylated genes have been identified as potential markers for 5-FU resistance, such as MHL1[240], uridine-cytidine kinase 2 (UCK2 or UMPK)[241] death-associated protein kinase (DAPK)[242], as well as MGMT[243] and XAF1[244] (antagonizes the anticaspase activity of X-linked inhibitor of apoptosis and may be important in mediating apoptosis resistance in cancer cells). The *in vitro* data generated for this thesis from cell lines treated with 5-FU, oxaliplatin and irinotecan confirmed the observation that DKK4 overexpression leads to increased 5-FU chemoresistance in CRC cell lines, while introduction of TFAP2E was associated with an increased sensitivity towards 5-FU treatment. This *in vitro* observation was further supported by the analysis of biopsies from

220 patients in 4 cohorts undergoing 5-FU based chemotherapy or chemoradiation in colon or rectal cancers. We found a strong association between TFAP2E methylation and lack of chemotherapy response in the tumor. Fixed effect and random model analysis of the pooled cohort data revealed an estimated 5-fold higher response probability for non-methylated patients. Interestingly, the correlation was observed in primary rectal cancers and metastatic colorectal cancers, independent of the treatment with 5-FU based chemotherapy or chemoradiation. Also, assessment of response in these cohorts, either by standard RECIST criteria or by histological response, did not influence this strong association, indicating that TFAP2E methylation may be valuable for response prediction in either setting.

The TUSC3 gene is located on 8p22, a chromosomal region which is either homozogously or heterozogously deleted (LOH) for example in lung[245], prostate[246, 247] and breast cancer[248] and was found to be silenced in various epithelial (including colorectal) cancer cell lines by hypermethylation.[249] Hypermethylation of TUSC3 was also found in glioblastoma, in concordance with estrogen receptor methylation[250] and prostate cancer.[251] The gene is expressed in most nonlymphoid human tissues including prostate, lung, liver, and colon. Expression was also detected in many epithelial tumor cell lines. Loss of TUSC3 expression seems to have an impact on survival (in ovarian cancer)[252] and development of lymph node metastasis[253, 254], at least in some epithelial cancer types. The TUSC3 protein contains an oligosaccharyltransferase (OST) domain, which shares a 20% identity with the yeast OST3 subunit and several transmembrane domains, which suggests that it is membrane bound. Little is known about the potential functions of this gene, a defect through homozygous deletion was recently linked to be associated with mental retardation in two studies, one german[255] and one french.[256] This was thought to be caused by a defect in the protein N-glycosylation process (through the assumed OST activity of the gene), however analyses of patient fibroblasts showed normal N-glycan synthesis and transfer. A recent study suggests that TUSC3 rather plays a role in cellular magnesium uptake, since knockdown significantly lowers the total and free intracellular magnesium $Mg(2+)$ concentrations in mammalian cell lines[257] through the plasma membrane and is important in embryonic development in mammals.

Two transcript variants encoding distinct isoforms have been identified for TUSC3, the longer one contains a redox active Thioredoxin domain, whereas in the shorter on this is reduced to a Thioredoxin like domain. This suggests TUSC3 could be located in the membrane of the endoplasmic reticulum. There is also evidence for association of TUSC3 methylation with preeclampsia and it is possible it represents a polymorphically imprinted gene[258] on the maternal allele. In CRC, methylation levels of TUSC3 also correlate with age.[259, 260] However, this seems to

be the cause of a field defect, which affects a group of aberrantly methylated genes in colon mucosa.[261] No link between TUSC3 methylation and APC/WNT, RAS/RAF, or P53 pathways was found in colorectal carcinoma[262] so far. However, in this study a significant correlation between TUSC3 methylation and a combination of APC and TP53 mutations was observed in a group of 63 patients. This marks a novel finding which should be validated in other patient cohorts and evaluated in detail. The observed rate of hypermethylation at the TUSC3 promoter points out to two things, first that TUSC3 methylation is occurring in a subset of CRC patients (around 30-40%) which would be expected from a single marker, and second, that TUSC3 methylation occurs in normal epithelial tissue as well, possible as an age related methylation field defect. The fact that no correlation with patient age was detected in this study (see results section) is probably due to the fact that almost all patients are over 50 years of age (mean age 67.5 years, see appendix A for details). While the fact that TUSC3 mRNA expression was detected in all tumors and normal mucosa might seems puzzling at first, but can be explained by the fact that some cell types like fibroblasts might not be affected through methylation in contrast to the epithelial cells (both tumor and nontumor). The protein expression levels found in the tumors fit to the found degree of methylation, pointing again to a subset of TUSC3 negative tumors (around 30%). These tumors seem to be characterized by a significant worse prognosis, indicating that TUSC3 is indeed a tumor suppressor gene. The fact that these observation is depending on lymph node invasion suggests that TUSC3 plays a role in early events of tumorigenesis and can not compensate for later occurring events (or "hits"). This fits to the finding that the key target for TUSC3 is SPARC, for protein for which a similar observation regarding expression and prognosis in CRC has been made in a set of 292 primary colorectal cancers by immunostaining.[263] The here confirmed TUSC3 downstream target SPARC (also known as osteonectin) is a matrix-associated glycoprotein, and is the most common noncollagenous protein in bone. SPARC adheres collagen and calcium and influences the synthesis of extracellular matrix (ECM); however, the role of SPARC in tissues has not been clearly elucidated.[264] Many diverse functions have been reported, including: interaction with the ECM in epithelial cells and stromal cells; cell motility; cell adhesion; tumor invasion; and proliferation of normal and tumor cells.[263] The loss of SPARC protein expression occurs also in a subset of tumors (also around 30%) and the survival rates for SPARC positive CRC patients are very similar to the observed survival rates for TUSC3 positive patients (75.83% vs. 56.79% for SPARC positive and SPARC negative versus 83.5% for TUSC3 positive and 68.6% for TUSC3 negative), if lymph node status is taken into account for TUSC3. While for SPARC expression and lymph node status no survival data in CRC is available so far, the correlation with lymph node status has also been observed in gastric cancer[265], especially of the intestinal type [266] but both studies had only a low number of patients with survival data (58 and 43 patients). The fact that SPARC is mainly expressed in stromal cells and not the tumor cells itself (where it is silenced through methylation) also fits to the fact that TUSC3 expression could be detected in tumor and nontumor tissue despite being methylated, leading to the assumption that stromal cells like fibroblasts are not affected by TUSC3 methylation. Downstream of TUSC3 three other genes besides SPARC could be confirmed as potential targets: HSPA6 which encodes Heat Shock Protein 70 is associated with stress response, COL6A3 (collagen, type VI, alpha 3) and NUPR1 (nuclear protein, transcriptional regulator, 1) are stroma markers like SPARC.[267] While HSPA6 links TUSC3 to stress, the effect on COL6A3 and NUPR1 expression could be explained as a side effect of SPARC, especially for COL6A3 since SPARC is able to bind to collagen.[268] SPARC has also been linked to adhesion and migration in several cancers, including colorectal cancer.[269] For breast cancer, it has shown that high SPARC expression leads to decreased in vitro invasion of osteonectin-infected cells through Matrigel and colony formation on Matrigel and inhibited metastasis in a dose-dependent manner to many different organs including bone.[270] In our case, a similar effect on adhesion and migration with collagen coated surfaces was observed for TUSC3 transfected cells. In summary, we show that TUSC3 expression is lost in a subset of approximately on third of colorectal tumors. These tumors are associated with worse prognosis and lymph node invasion, which is due to the loss of SPARC expression in turn. The worse prognosis is due to a higher risk for metastasis since tumor due the effects of SPARC on adhesion and migration. Since SPARC is also associated with therapy resistance[271], tumors in these subgroup could be treated with 5-Azacytidine to reinduce TUSC3 and SPARC expression and enhance chemotherapy.[272]

The relaxin/insulin-like family peptide receptor 3 is a gene protein coupled receptor belonging to a subclass of four closely related G protein-coupled receptors that bind relaxin peptide hormones; Relaxin 3 is the ligand for this particular receptor. Both the receptor and its ligand are highly conserved among mammals and localization of RXFP3 in the brain of primates and rodents seem to be comparable.[273] Activation of RXFP3 leads to activation of MAPK1/2 (mitogen-activated protein kinases, MAPKs, also known as extracellular signal-regulated kinases, ERKs, act as an integration point for multiple biochemical signals, and are involved in a wide variety of cellular processes such as proliferation, differentiation, transcription regulation and development) signaling.[274] RXFP3 is expressed in a wide range of tissues, such as brain, testis, thymus, adrenal glands, pancreas, mammary glands, placenta, salivary glands, spinal cord and stomach.[275] The mRNA consists of only one exon and no variants are known so far, but the protein has several membrane domains. Not

much is known about the role of this gene outside of the neurological context. In mammalian brains, RXFP3 modulation has effects on feeding and metabolism, the activity of the septohippocampal pathway, and spatial memory[276] as well as stress response in the mouse brain.[277] Hypermethylation of RXFP3 was found in endometrial tumors and associated with microsatellite instability and loss of expression in these tumors was inversely associated with promoter hypermethylation.[278] However, RXFP3 hypermethylation was significantly correlated with disease-free survival in this study. Here we report hypermethylation of RXFP3 in over 50% of colorectal tumors in a cohort of 74 patients and no age related methylation in corresponding normal tissue, thus pointing out the potential value of RXFP3 as a diagnostic tool for colorectal cancer. Furthermore, RXFP3 methylation was associated with a combination of KRAS and APC mutations in those tumors, not only linking RXFP3 (probably through its activation of MAPK signaling) to TGF beta (a cytokine and multifunctional protein that regulate proliferation, differentiation, adhesion, migration, and other functions in many cell types) and WNT signaling, but also providing more hints to the involvement of RXFP3 hypermethylation in early events of tumor formation (since KRAS and APC could be seen as gatekeeper mutations in tumor development). Since most of the identified target genes which could be validated via qPCR were factors for hypoxia, this links RXFP3 to stress response not only in the central nervous system but also for the first time in the gastrointestinal tract.

To sum up, the goal of this thesis was the evaluation and functional characterisation of a couple of DNA methylation markers from a given set as useful biomarkers for clinical relevant applications including diagnosis, prognosis and response prediction in colorectal cancer. However, further research is needed to put the new gained insights into clinical practice. To illustrate the usefulness of DNA promoter hypermethylation of potential tumor suppressor genes as biomarkers for clinical applications, three single markers were finally analyzed as an example for the three aforementioned main purposes in a clinical setting.

6. References

1. Jemal A, Bray F, Center MM, Ferlay J, Ward E, Forman D. Global cancer statistics. CA Cancer J Clin 2011;61:69-90.
2. Center MM, Jemal A, Smith RA, Ward E. Worldwide variations in colorectal cancer. CA Cancer J Clin 2009;59:366-78.
3. Center MM, Jemal A, Ward E. International trends in colorectal cancer incidence rates. Cancer Epidemiol Biomarkers Prev 2009;18:1688-94.
4. Jemal A, Siegel R, Xu J, Ward E. Cancer statistics, 2010. CA Cancer J Clin;60:277-300.
5. Siegel RL, Jemal A, Ward EM. Increase in incidence of colorectal cancer among young men and women in the United States. Cancer Epidemiol Biomarkers Prev 2009;18:1695-8.
6. Ferlay J, Parkin DM, Steliarova-Foucher E. Estimates of cancer incidence and mortality in Europe in 2008. Eur J Cancer;46:765-81.
7. Haberland J, Bertz J, Wolf U, Ziese T, Kurth BM. German cancer statistics 2004. BMC Cancer;10:52.
8. Treanor D, Quirke P. Pathology of colorectal cancer. Clin Oncol (R Coll Radiol) 2007;19:769-76.
9. Moslein G. [Hereditary colorectal cancer]. Chirurg 2008;79:1038-46.
10. Cappell MS. The pathophysiology, clinical presentation, and diagnosis of colon cancer and adenomatous polyps. Med Clin North Am 2005;89:1-42, vii.
11. Fearon ER. Molecular genetics of colorectal cancer. Annu Rev Pathol 2011;6:479-507.
12. Feinberg AP, Ohlsson R, Henikoff S. The epigenetic progenitor origin of human cancer. Nat Rev Genet 2006;7:21-33.
13. Davies RJ, Miller R, Coleman N. Colorectal cancer screening: prospects for molecular stool analysis. Nat Rev Cancer 2005;5:199-209.
14. Schmiegel W, Pox C, Arnold D, Porschen R, Rodel C, Reinacher-Schick A. Colorectal carcinoma: the management of polyps, (neo)adjuvant therapy, and the treatment of metastases. Dtsch Arztebl Int 2009;106:843-8.
15. Markowitz SD, Bertagnolli MM. Molecular origins of cancer: Molecular basis of colorectal cancer. N Engl J Med 2009;361:2449-60.
16. Castells A, Paya A, Alenda C, et al. Cyclooxygenase 2 expression in colorectal cancer with DNA mismatch repair deficiency. Clin Cancer Res 2006;12:1686-92.
17. Fearon ER, Vogelstein B. A genetic model for colorectal tumorigenesis. Cell 1990;61:759-67.
18. Power DG, Lichtman SM. Chemotherapy for the elderly patient with colorectal cancer. Cancer J 2010;16:241-52.
19. Mai PL, Wideroff L, Greene MH, Graubard BI. Prevalence of family history of breast, colorectal, prostate, and lung cancer in a population-based study. Public Health Genomics 2010;13:495-503.
20. Stein QP, Flanagan JD. Genetic and familial factors influencing breast, colon, prostate and lung cancers. S D Med 2010;Spec No:16-22.
21. Liang PS, Chen TY, Giovannucci E. Cigarette smoking and colorectal cancer incidence and mortality: systematic review and meta-analysis. Int J Cancer 2009;124:2406-15.
22. Chan AT, Giovannucci EL. Primary prevention of colorectal cancer. Gastroenterology 2010;138:2029-43 e10.
23. Bujanda L, Cosme A, Gil I, Arenas-Mirave JI. Malignant colorectal polyps. World J Gastroenterol 2010;16:3103-11.
24. Bond JH. Colon polyps and cancer. Endoscopy 2005;37:208-12.

References

25. May A, Ell C. [Gastroenterological endoscopy]. Dtsch Med Wochenschr 2011;136:206-8.
26. Ullman TA, Itzkowitz SH. Intestinal inflammation and cancer. Gastroenterology 2011;140:1807-16 e1.
27. Khan N, Afaq F, Mukhtar H. Lifestyle as risk factor for cancer: Evidence from human studies. Cancer Lett 2010;293:133-43.
28. Hasan N, Pollack A, Cho I. Infectious causes of colorectal cancer. Infect Dis Clin North Am 2010;24:1019-39, x.
29. Key TJ. Fruit and vegetables and cancer risk. Br J Cancer 2010;104:6-11.
30. Larsson SC, Orsini N, Wolk A. Vitamin B6 and risk of colorectal cancer: a meta-analysis of prospective studies. Jama 2010;303:1077-83.
31. Sung MK, Bae YJ. Linking obesity to colorectal cancer: application of nutrigenomics. Biotechnol J 2010;5:930-41.
32. Lorenzon L, Ferri M, Pilozzi E, Torrisi MR, Ziparo V, French D. Human papillomavirus and colorectal cancer: evidences and pitfalls of published literature. Int J Colorectal Dis 2011;26:135-42.
33. Young GP, Cole SR. Which fecal occult blood test is best to screen for colorectal cancer? Nat Clin Pract Gastroenterol Hepatol 2009;6:140-1.
34. Imperiale TF, Ransohoff DF, Itzkowitz SH, Turnbull BA, Ross ME. Fecal DNA versus fecal occult blood for colorectal-cancer screening in an average-risk population. N Engl J Med 2004;351:2704-14.
35. Grazzini G, Visioli CB, Zorzi M, et al. Immunochemical faecal occult blood test: number of samples and positivity cutoff. What is the best strategy for colorectal cancer screening? Br J Cancer 2009;100:259-65.
36. van Rossum LG, van Rijn AF, van Oijen MG, et al. False negative fecal occult blood tests due to delayed sample return in colorectal cancer screening. Int J Cancer 2009;125:746-50.
37. Sieg A, Theilmeier A. [Results of coloscopy screening in 2005--an Internet-based documentation]. Dtsch Med Wochenschr 2006;131:379-83.
38. Atkin WS, Edwards R, Kralj-Hans I, et al. Once-only flexible sigmoidoscopy screening in prevention of colorectal cancer: a multicentre randomised controlled trial. Lancet 2010;375:1624-33.
39. Greenwald B. The stool DNA test: an emerging technology in colorectal cancer screening. Gastroenterol Nurs 2005;28:28-32.
40. Young GP, Bosch LJ. Fecal Tests: From Blood to Molecular Markers. Curr Colorectal Cancer Rep 2011;7:62-70.
41. Geiger TM, Ricciardi R. Screening options and recommendations for colorectal cancer. Clin Colon Rectal Surg 2009;22:209-17.
42. Diehl F, Schmidt K, Choti MA, et al. Circulating mutant DNA to assess tumor dynamics. Nat Med 2008;14:985-90.
43. deVos T, Tetzner R, Model F, et al. Circulating methylated SEPT9 DNA in plasma is a biomarker for colorectal cancer. Clin Chem 2009;55:1337-46.
44. Puppa G, Sonzogni A, Colombari R, Pelosi G. TNM staging system of colorectal carcinoma: a critical appraisal of challenging issues. Arch Pathol Lab Med 2010;134:837-52.
45. Quirke P, Cuvelier C, Ensari A, et al. Evidence-based medicine: the time has come to set standards for staging. J Pathol 2010;221:357-60.
46. Edge SB, Compton CC. The American Joint Committee on Cancer: the 7th edition of the AJCC cancer staging manual and the future of TNM. Ann Surg Oncol 2010;17:1471-4.
47. Kyriakos M. The President's cancer, the Dukes classification, and confusion. Arch Pathol Lab Med 1985;109:1063-6.

48. Bosman FT. Prognostic value of pathological characteristics of colorectal cancer. Eur J Cancer 1995;31A:1216-21.
49. Ades S. Adjuvant chemotherapy for colon cancer in the elderly: moving from evidence to practice. Oncology (Williston Park) 2009;23:162-7.
50. Ruiz-Tovar J, Jimenez-Miramon J, Valle A, Limones M. Endoscopic resection as unique treatment for early colorectal cancer. Rev Esp Enferm Dig 2010;102:435-41.
51. Okuno K. Surgical treatment for digestive cancer. Current issues - colon cancer. Dig Surg 2007;24:108-14.
52. Neumann UP, Seehofer D, Neuhaus P. The surgical treatment of hepatic metastases in colorectal carcinoma. Dtsch Arztebl Int 2010;107:335-42.
53. Rodel C, Knoefel WT, Schlitt HJ, Staib L, Hohler T. [Neoadjuvant and surgical treatment for rectal cancer]. Onkologie 2009;32 Suppl 2:17-20.
54. Aschele C, Bergamo F, Lonardi S. Chemotherapy for operable and advanced colorectal cancer. Cancer Treat Rev 2009;35:509-16.
55. Van Loon K, Venook AP. Adjuvant treatment of colon cancer: what is next? Curr Opin Oncol 2011.
56. Van Cutsem E, Kohne CH, Hitre E, et al. Cetuximab and chemotherapy as initial treatment for metastatic colorectal cancer. N Engl J Med 2009;360:1408-17.
57. Raftery L, Goldberg RM. Optimal delivery of cytotoxic chemotherapy for colon cancer. Cancer J 2011;16:214-9.
58. Gollins S. Radiation, chemotherapy and biological therapy in the curative treatment of locally advanced rectal cancer. Colorectal Dis 2010;12 Suppl 2:2-24.
59. Minsky BD. Chemoradiation for rectal cancer: rationale, approaches, and controversies. Surg Oncol Clin N Am 2010;19:803-18.
60. Julien LA, Thorson AG. Current neoadjuvant strategies in rectal cancer. J Surg Oncol 2010;101:321-6.
61. Platell C. Colorectal cancer survival. ANZ J Surg 2011;81:310-1.
62. Rees M, Tekkis PP, Welsh FK, O'Rourke T, John TG. Evaluation of long-term survival after hepatic resection for metastatic colorectal cancer: a multifactorial model of 929 patients. Ann Surg 2008;247:125-35.
63. Gallagher DJ, Kemeny N. Metastatic colorectal cancer: from improved survival to potential cure. Oncology 2011;78:237-48.
64. Brown RE, Bower MR, Martin RC. Hepatic resection for colorectal liver metastases. Surg Clin North Am 2011;90:839-52.
65. Lieberman D. Progress and challenges in colorectal cancer screening and surveillance. Gastroenterology 2010;138:2115-26.
66. Smith RA, Cokkinides V, Brooks D, Saslow D, Brawley OW. Cancer screening in the United States, 2010: a review of current American Cancer Society guidelines and issues in cancer screening. CA Cancer J Clin 2010;60:99-119.
67. Haggstrom DA, Imperiale TF. Surveillance approaches among colorectal cancer survivors after curative-intent. Minerva Gastroenterol Dietol 2009;55:483-500.
68. Van Cutsem E, Oliveira J. Primary colon cancer: ESMO clinical recommendations for diagnosis, adjuvant treatment and follow-up. Ann Oncol 2009;20 Suppl 4:49-50.
69. Glimelius B, Oliveira J. Rectal cancer: ESMO clinical recommendations for diagnosis, treatment and follow-up. Ann Oncol 2009;20 Suppl 4:54-6.
70. Bird A. Perceptions of epigenetics. Nature 2007;447:396-8.
71. Jones PA, Baylin SB. The epigenomics of cancer. Cell 2007;128:683-92.
72. Galm O, Herman JG, Baylin SB. The fundamental role of epigenetics in hematopoietic malignancies. Blood Rev 2006;20:1-13.

73. Sharma S, Kelly TK, Jones PA. Epigenetics in cancer. Carcinogenesis 2010;31:27-36.
74. Miranda TB, Jones PA. DNA methylation: the nuts and bolts of repression. J Cell Physiol 2007;213:384-90.
75. Kim JK, Samaranayake M, Pradhan S. Epigenetic mechanisms in mammals. Cell Mol Life Sci 2009;66:596-612.
76. Barros SP, Offenbacher S. Epigenetics: connecting environment and genotype to phenotype and disease. J Dent Res 2009;88:400-8.
77. Gronbaek K, Hother C, Jones PA. Epigenetic changes in cancer. Apmis 2007;115:1039-59.
78. Bernstein BE, Meissner A, Lander ES. The mammalian epigenome. Cell 2007;128:669-81.
79. Suzuki MM, Bird A. DNA methylation landscapes: provocative insights from epigenomics. Nat Rev Genet 2008;9:465-76.
80. Turner BM. Defining an epigenetic code. Nat Cell Biol 2007;9:2-6.
81. Mariman EC. Epigenetic manifestations in diet-related disorders. J Nutrigenet Nutrigenomics 2008;1:232-9.
82. Herceg Z. Epigenetic information in chromatin and cancer. Eur J Cancer 2009;45 Suppl 1:442-4.
83. Chahwan R, Wontakal SN, Roa S. The multidimensional nature of epigenetic information and its role in disease. Discov Med 2011;11:233-43.
84. Talbert PB, Henikoff S. Histone variants--ancient wrap artists of the epigenome. Nat Rev Mol Cell Biol 2011;11:264-75.
85. Iacobuzio-Donahue CA. Epigenetic changes in cancer. Annu Rev Pathol 2009;4:229-49.
86. Kristensen LS, Nielsen HM, Hansen LL. Epigenetics and cancer treatment. Eur J Pharmacol 2009;625:131-42.
87. Taby R, Issa JP. Cancer epigenetics. CA Cancer J Clin 2010;60:376-92.
88. Reik W. Stability and flexibility of epigenetic gene regulation in mammalian development. Nature 2007;447:425-32.
89. Ropero S, Esteller M. The role of histone deacetylases (HDACs) in human cancer. Mol Oncol 2007;1:19-25.
90. Lopez J, Percharde M, Coley HM, Webb A, Crook T. The context and potential of epigenetics in oncology. Br J Cancer 2009;100:571-7.
91. Portela A, Esteller M. Epigenetic modifications and human disease. Nat Biotechnol 2010;28:1057-68.
92. Lister R, Pelizzola M, Dowen RH, et al. Human DNA methylomes at base resolution show widespread epigenomic differences. Nature 2009;462:315-22.
93. Law JA, Jacobsen SE. Establishing, maintaining and modifying DNA methylation patterns in plants and animals. Nat Rev Genet 2010;11:204-20.
94. Jabbari K, Bernardi G. Cytosine methylation and CpG, TpG (CpA) and TpA frequencies. Gene 2004;333:143-9.
95. Daura-Oller E, Cabre M, Montero MA, Paternain JL, Romeu A. Specific gene hypomethylation and cancer: new insights into coding region feature trends. Bioinformation 2009;3:340-3.
96. Saxonov S, Berg P, Brutlag DL. A genome-wide analysis of CpG dinucleotides in the human genome distinguishes two distinct classes of promoters. Proc Natl Acad Sci U S A 2006;103:1412-7.
97. Takai D, Jones PA. Comprehensive analysis of CpG islands in human chromosomes 21 and 22. Proc Natl Acad Sci U S A 2002;99:3740-5.
98. Irizarry RA, Ladd-Acosta C, Wen B, et al. The human colon cancer methylome shows similar hypo- and hypermethylation at conserved tissue-specific CpG island shores. Nat Genet 2009;41:178-86.

99. Ji H, Ehrlich LI, Seita J, et al. Comprehensive methylome map of lineage commitment from haematopoietic progenitors. Nature 2010;467:338-42.
100. Feil R, Berger F. Convergent evolution of genomic imprinting in plants and mammals. Trends Genet 2007;23:192-9.
101. Hirasawa R, Feil R. Genomic imprinting and human disease. Essays Biochem 2010;48:187-200.
102. Watanabe Y, Maekawa M. Methylation of DNA in cancer. Adv Clin Chem 2010;52:145-67.
103. Konkel MK, Batzer MA. A mobile threat to genome stability: The impact of non-LTR retrotransposons upon the human genome. Semin Cancer Biol 2010;20:211-21.
104. Schulz WA, Steinhoff C, Florl AR. Methylation of endogenous human retroelements in health and disease. Curr Top Microbiol Immunol 2006;310:211-50.
105. Sulewska A, Niklinska W, Kozlowski M, et al. DNA methylation in states of cell physiology and pathology. Folia Histochem Cytobiol 2007;45:149-58.
106. Sansom OJ, Maddison K, Clarke AR. Mechanisms of disease: methyl-binding domain proteins as potential therapeutic targets in cancer. Nat Clin Pract Oncol 2007;4:305-15.
107. Sasai N, Defossez PA. Many paths to one goal? The proteins that recognize methylated DNA in eukaryotes. Int J Dev Biol 2009;53:323-34.
108. Hashimoto H, Vertino PM, Cheng X. Molecular coupling of DNA methylation and histone methylation. Epigenomics 2010;2:657-69.
109. Cheng X, Blumenthal RM. Coordinated chromatin control: structural and functional linkage of DNA and histone methylation. Biochemistry 2010;49:2999-3008.
110. Daniel FI, Cherubini K, Yurgel LS, de Figueiredo MA, Salum FG. The role of epigenetic transcription repression and DNA methyltransferases in cancer. Cancer 2010;117:677-87.
111. Teodoridis JM, Hardie C, Brown R. CpG island methylator phenotype (CIMP) in cancer: causes and implications. Cancer Lett 2008;268:177-86.
112. Biniszkiewicz D, Gribnau J, Ramsahoye B, et al. Dnmt1 overexpression causes genomic hypermethylation, loss of imprinting, and embryonic lethality. Mol Cell Biol 2002;22:2124-35.
113. Hirasawa R, Chiba H, Kaneda M, et al. Maternal and zygotic Dnmt1 are necessary and sufficient for the maintenance of DNA methylation imprints during preimplantation development. Genes Dev 2008;22:1607-16.
114. Damelin M, Bestor TH. Biological functions of DNA methyltransferase 1 require its methyltransferase activity. Mol Cell Biol 2007;27:3891-9.
115. Egger G, Jeong S, Escobar SG, et al. Identification of DNMT1 (DNA methyltransferase 1) hypomorphs in somatic knockouts suggests an essential role for DNMT1 in cell survival. Proc Natl Acad Sci U S A 2006;103:14080-5.
116. Kinney SR, Pradhan S. Regulation of expression and activity of DNA (Cytosine-5) methyltransferases in Mammalian cells. Prog Mol Biol Transl Sci 2011;101:311-33.
117. Chedin F. The DNMT3 Family of Mammalian De Novo DNA Methyltransferases. Prog Mol Biol Transl Sci 2011;101:255-85.
118. Goll MG, Kirpekar F, Maggert KA, et al. Methylation of tRNAAsp by the DNA methyltransferase homolog Dnmt2. Science 2006;311:395-8.
119. Thiagarajan D, Dev RR, Khosla S. The DNA methyltranferase Dnmt2 participates in RNA processing during cellular stress. Epigenetics 2010;6:103-13.
120. Jurkowski TP, Meusburger M, Phalke S, et al. Human DNMT2 methylates tRNA(Asp) molecules using a DNA methyltransferase-like catalytic mechanism. Rna 2008;14:1663-70.
121. Goel A, Boland CR. Recent insights into the pathogenesis of colorectal cancer. Curr Opin Gastroenterol 2010;26:47-52.
122. Veeck J, Esteller M. Breast cancer epigenetics: from DNA methylation to microRNAs. J Mammary Gland Biol Neoplasia 2010;15:5-17.

123. Ng EK, Tsang WP, Ng SS, et al. MicroRNA-143 targets DNA methyltransferases 3A in colorectal cancer. Br J Cancer 2009;101:699-706.
124. Garzon R, Liu S, Fabbri M, et al. MicroRNA-29b induces global DNA hypomethylation and tumor suppressor gene reexpression in acute myeloid leukemia by targeting directly DNMT3A and 3B and indirectly DNMT1. Blood 2009;113:6411-8.
125. Fritz EL, Papavasiliou FN. Cytidine deaminases: AIDing DNA demethylation? Genes Dev 2010;24:2107-14.
126. Mohr F, Dohner K, Buske C, Rawat VP. TET genes: new players in DNA demethylation and important determinants for stemness. Exp Hematol 2011;39:272-81.
127. Wu SC, Zhang Y. Active DNA demethylation: many roads lead to Rome. Nat Rev Mol Cell Biol 2010;11:607-20.
128. Guil S, Esteller M. DNA methylomes, histone codes and miRNAs: tying it all together. Int J Biochem Cell Biol 2009;41:87-95.
129. Iorio MV, Piovan C, Croce CM. Interplay between microRNAs and the epigenetic machinery: an intricate network. Biochim Biophys Acta 2010;1799:694-701.
130. Breving K, Esquela-Kerscher A. The complexities of microRNA regulation: mirandering around the rules. Int J Biochem Cell Biol 2010;42:1316-29.
131. Friedman JM, Liang G, Liu CC, et al. The putative tumor suppressor microRNA-101 modulates the cancer epigenome by repressing the polycomb group protein EZH2. Cancer Res 2009;69:2623-9.
132. Noonan EJ, Place RF, Pookot D, et al. miR-449a targets HDAC-1 and induces growth arrest in prostate cancer. Oncogene 2009;28:1714-24.
133. Tuddenham L, Wheeler G, Ntounia-Fousara S, et al. The cartilage specific microRNA-140 targets histone deacetylase 4 in mouse cells. FEBS Lett 2006;580:4214-7.
134. Sinkkonen L, Hugenschmidt T, Berninger P, et al. MicroRNAs control de novo DNA methylation through regulation of transcriptional repressors in mouse embryonic stem cells. Nat Struct Mol Biol 2008;15:259-67.
135. Benetti R, Gonzalo S, Jaco I, et al. A mammalian microRNA cluster controls DNA methylation and telomere recombination via Rbl2-dependent regulation of DNA methyltransferases. Nat Struct Mol Biol 2008;15:268-79.
136. Davalos V, Esteller M. MicroRNAs and cancer epigenetics: a macrorevolution. Curr Opin Oncol 2010;22:35-45.
137. Carmona FJ, Esteller M. Epigenomics of human colon cancer. Mutat Res 2010.
138. Ushijima T. Epigenetic field for cancerization. J Biochem Mol Biol 2007;40:142-50.
139. Esteller M. Epigenetics in cancer. N Engl J Med 2008;358:1148-59.
140. Kaneda A, Yagi K. Two groups of DNA methylation markers to classify colorectal cancer into three epigenotypes. Cancer Sci 2011;102:18-24.
141. Kanwal R, Gupta S. Epigenetics and cancer. J Appl Physiol 2010;109:598-605.
142. Albert M, Helin K. Histone methyltransferases in cancer. Semin Cell Dev Biol 2010;21:209-20.
143. Ganesan A, Nolan L, Crabb SJ, Packham G. Epigenetic therapy: histone acetylation, DNA methylation and anti-cancer drug discovery. Curr Cancer Drug Targets 2009;9:963-81.
144. Mai A, Altucci L. Epi-drugs to fight cancer: from chemistry to cancer treatment, the road ahead. Int J Biochem Cell Biol 2009;41:199-213.
145. Copeland RA, Olhava EJ, Scott MP. Targeting epigenetic enzymes for drug discovery. Curr Opin Chem Biol 2010;14:505-10.
146. Issa JP, Kantarjian HM. Targeting DNA methylation. Clin Cancer Res 2009;15:3938-46.
147. Grant S, Easley C, Kirkpatrick P. Vorinostat. Nat Rev Drug Discov 2007;6:21-2.

148. Witt O, Deubzer HE, Milde T, Oehme I. HDAC family: What are the cancer relevant targets? Cancer Lett 2009;277:8-21.
149. Prince HM, Bishton MJ, Harrison SJ. Clinical studies of histone deacetylase inhibitors. Clin Cancer Res 2009;15:3958-69.
150. Witt O, Deubzer HE, Lodrini M, Milde T, Oehme I. Targeting histone deacetylases in neuroblastoma. Curr Pharm Des 2009;15:436-47.
151. Tan J, Cang S, Ma Y, Petrillo RL, Liu D. Novel histone deacetylase inhibitors in clinical trials as anti-cancer agents. J Hematol Oncol 2010;3:5.
152. Cortez CC, Jones PA. Chromatin, cancer and drug therapies. Mutat Res 2008;647:44-51.
153. Carew JS, Giles FJ, Nawrocki ST. Histone deacetylase inhibitors: mechanisms of cell death and promise in combination cancer therapy. Cancer Lett 2008;269:7-17.
154. Steele N, Finn P, Brown R, Plumb JA. Combined inhibition of DNA methylation and histone acetylation enhances gene re-expression and drug sensitivity in vivo. Br J Cancer 2009;100:758-63.
155. Ramalingam SS, Belani CP, Ruel C, et al. Phase II study of belinostat (PXD101), a histone deacetylase inhibitor, for second line therapy of advanced malignant pleural mesothelioma. J Thorac Oncol 2009;4:97-101.
156. Owonikoko TK, Ramalingam SS, Kanterewicz B, Balius TE, Belani CP, Hershberger PA. Vorinostat increases carboplatin and paclitaxel activity in non-small-cell lung cancer cells. Int J Cancer 2010;126:743-55.
157. Weiss JF, Landauer MR. History and development of radiation-protective agents. Int J Radiat Biol 2009;85:539-73.
158. Miranda TB, Cortez CC, Yoo CB, et al. DZNep is a global histone methylation inhibitor that reactivates developmental genes not silenced by DNA methylation. Mol Cancer Ther 2009;8:1579-88.
159. Blum W, Garzon R, Klisovic RB, et al. Clinical response and miR-29b predictive significance in older AML patients treated with a 10-day schedule of decitabine. Proc Natl Acad Sci U S A 2010;107:7473-8.
160. Kristensen LS, Hansen LL. PCR-based methods for detecting single-locus DNA methylation biomarkers in cancer diagnostics, prognostics, and response to treatment. Clin Chem 2009;55:1471-83.
161. Kristensen LS, Wojdacz TK, Thestrup BB, Wiuf C, Hager H, Hansen LL. Quality assessment of DNA derived from up to 30 years old formalin fixed paraffin embedded (FFPE) tissue for PCR-based methylation analysis using SMART-MSP and MS-HRM. BMC Cancer 2009;9:453.
162. Levenson VV. DNA methylation as a universal biomarker. Expert Rev Mol Diagn 2010;10:481-8.
163. Kim MS, Lee J, Sidransky D. DNA methylation markers in colorectal cancer. Cancer Metastasis Rev 2010;29:181-206.
164. Cheung HH, Lee TL, Rennert OM, Chan WY. DNA methylation of cancer genome. Birth Defects Res C Embryo Today 2009;87:335-50.
165. Model F, Osborn N, Ahlquist D, et al. Identification and validation of colorectal neoplasia-specific methylation markers for accurate classification of disease. Mol Cancer Res 2007;5:153-63.
166. Lofton-Day C, Model F, Devos T, et al. DNA methylation biomarkers for blood-based colorectal cancer screening. Clin Chem 2008;54:414-23.
167. Payne SR, Serth J, Schostak M, et al. DNA methylation biomarkers of prostate cancer: Confirmation of candidates and evidence urine is the most sensitive body fluid for non-invasive detection. Prostate 2009.

References

168. Kunitz A, Wolter M, van den Boom J, et al. DNA hypermethylation and aberrant expression of the EMP3 gene at 19q13.3 in Human Gliomas. Brain Pathol 2007;17:363-70.
169. Weller M, Stupp R, Reifenberger G, et al. MGMT promoter methylation in malignant gliomas: ready for personalized medicine? Nat Rev Neurol 2010;6:39-51.
170. Eads CA, Danenberg KD, Kawakami K, et al. MethyLight: a high-throughput assay to measure DNA methylation. Nucleic Acids Res 2000;28:E32.
171. Ebert MP, Model F, Mooney S, et al. Aristaless-like homeobox-4 gene methylation is a potential marker for colorectal adenocarcinomas. Gastroenterology 2006;131:1418-30.
172. Wojdacz TK, Dobrovic A, Hansen LL. Methylation-sensitive high-resolution melting. Nat Protoc 2008;3:1903-8.
173. Trinh BN, Long TI, Laird PW. DNA methylation analysis by MethyLight technology. Methods 2001;25:456-62.
174. Baehs S, Herbst A, Thieme SE, et al. Dickkopf-4 is frequently down-regulated and inhibits growth of colorectal cancer cells. Cancer Lett 2009;276:152-9.
175. Bax L, Yu LM, Ikeda N, Tsuruta H, Moons KG. Development and validation of MIX: comprehensive free software for meta-analysis of causal research data. BMC Med Res Methodol 2006;6:50.
176. Bax L, Yu LM, Ikeda N, Moons KG. A systematic comparison of software dedicated to meta-analysis of causal studies. BMC Med Res Methodol 2007;7:40.
177. Zhu J, Yao X. Use of DNA methylation for cancer detection: promises and challenges. Int J Biochem Cell Biol 2009;41:147-54.
178. Fassbender A, Lewin J, Konig T, et al. Quantitative DNA methylation profiling on a high-density oligonucleotide microarray. Methods Mol Biol 2010;576:155-70.
179. Deatherage DE, Potter D, Yan PS, Huang TH, Lin S. Methylation analysis by microarray. Methods Mol Biol 2009;556:117-39.
180. Walther A, Johnstone E, Swanton C, Midgley R, Tomlinson I, Kerr D. Genetic prognostic and predictive markers in colorectal cancer. Nat Rev Cancer 2009;9:489-99.
181. Lassmann S, Tang L, Capanu M, et al. Predictive molecular markers for colorectal cancer patients with resected liver metastasis and adjuvant chemotherapy. Gastroenterology 2007;133:1831-9.
182. Kweekel DM, Antonini NF, Nortier JW, Punt CJ, Gelderblom H, Guchelaar HJ. Explorative study to identify novel candidate genes related to oxaliplatin efficacy and toxicity using a DNA repair array. Br J Cancer 2009;101:357-62.
183. Crea F, Giovannetti E, Cortesi F, et al. Epigenetic mechanisms of irinotecan sensitivity in colorectal cancer cell lines. Mol Cancer Ther 2009;8:1964-73.
184. Ooyama A, Takechi T, Toda E, et al. Gene expression analysis using human cancer xenografts to identify novel predictive marker genes for the efficacy of 5-fluorouracil-based drugs. Cancer Sci 2006;97:510-22.
185. Brody JR, Hucl T, Costantino CL, et al. Limits to thymidylate synthase and TP53 genes as predictive determinants for fluoropyrimidine sensitivity and further evidence for RNA-based toxicity as a major influence. Cancer Res 2009;69:984-91.
186. Almendro V, Ametller E, Garcia-Recio S, et al. The role of MMP7 and its cross-talk with the FAS/FASL system during the acquisition of chemoresistance to oxaliplatin. PLoS One 2009;4:e4728.
187. Vie N, Copois V, Bascoul-Mollevi C, et al. Overexpression of phosphoserine aminotransferase PSAT1 stimulates cell growth and increases chemoresistance of colon cancer cells. Mol Cancer 2008;7:14.
188. Iwatsuki M, Mimori K, Yokobori T, et al. A platinum agent resistance gene, POLB, is a prognostic indicator in colorectal cancer. J Surg Oncol 2009.

189. Allen WL, Coyle VM, Jithesh PV, et al. Clinical determinants of response to irinotecan-based therapy derived from cell line models. Clin Cancer Res 2008;14:6647-55.
190. Gongora C, Candeil L, Vezzio N, et al. Altered expression of cell proliferation-related and interferon-stimulated genes in colon cancer cells resistant to SN38. Cancer Biol Ther 2008;7:822-32.
191. Braun AH, Stark K, Dirsch O, Hilger RA, Seeber S, Vanhoefer U. The epidermal growth factor receptor tyrosine kinase inhibitor gefitinib sensitizes colon cancer cells to irinotecan. Anticancer Drugs 2005;16:1099-108.
192. Candeil L, Gourdier I, Peyron D, et al. ABCG2 overexpression in colon cancer cells resistant to SN38 and in irinotecan-treated metastases. Int J Cancer 2004;109:848-54.
193. Moser C, Lang SA, Kainz S, et al. Blocking heat shock protein-90 inhibits the invasive properties and hepatic growth of human colon cancer cells and improves the efficacy of oxaliplatin in p53-deficient colon cancer tumors in vivo. Mol Cancer Ther 2007;6:2868-78.
194. Toscano F, Fajoui ZE, Gay F, et al. P53-mediated upregulation of DcR1 impairs oxaliplatin/TRAIL-induced synergistic anti-tumour potential in colon cancer cells. Oncogene 2008;27:4161-71.
195. Lee W, Belkhiri A, Lockhart AC, et al. Overexpression of OATP1B3 confers apoptotic resistance in colon cancer. Cancer Res 2008;68:10315-23.
196. Sinicrope FA, Rego RL, Halling KC, et al. Thymidylate synthase expression in colon carcinomas with microsatellite instability. Clin Cancer Res 2006;12:2738-44.
197. Sinicrope FA, Sargent DJ. Clinical implications of microsatellite instability in sporadic colon cancers. Curr Opin Oncol 2009;21:369-73.
198. Jensen SA, Vainer B, Kruhoffer M, Sorensen JB. Microsatellite instability in colorectal cancer and association with thymidylate synthase and dihydropyrimidine dehydrogenase expression. BMC Cancer 2009;9:25.
199. Iacopetta B, Kawakami K, Watanabe T. Predicting clinical outcome of 5-fluorouracil-based chemotherapy for colon cancer patients: is the CpG island methylator phenotype the 5-fluorouracil-responsive subgroup? Int J Clin Oncol 2008;13:498-503.
200. Tummala R, Romano RA, Fuchs E, Sinha S. Molecular cloning and characterization of AP-2 epsilon, a fifth member of the AP-2 family. Gene 2003;321:93-102.
201. Giaretti W, Molinu S, Ceccarelli J, Prevosto C. Chromosomal instability, aneuploidy, and gene mutations in human sporadic colorectal adenomas. Cell Oncol 2004;26:301-5.
202. Eckert D, Buhl S, Weber S, Jager R, Schorle H. The AP-2 family of transcription factors. Genome Biol 2005;6:246.
203. Orso F, Penna E, Cimino D, et al. AP-2alpha and AP-2gamma regulate tumor progression via specific genetic programs. Faseb J 2008;22:2702-14.
204. Orso F, Fassetta M, Penna E, et al. The AP-2alpha transcription factor regulates tumor cell migration and apoptosis. Adv Exp Med Biol 2007;604:87-95.
205. Schwartz B, Melnikova VO, Tellez C, et al. Loss of AP-2alpha results in deregulation of E-cadherin and MMP-9 and an increase in tumorigenicity of colon cancer cells in vivo. Oncogene 2007;26:4049-58.
206. Ruiz M, Pettaway C, Song R, Stoeltzing O, Ellis L, Bar-Eli M. Activator protein 2alpha inhibits tumorigenicity and represses vascular endothelial growth factor transcription in prostate cancer cells. Cancer Res 2004;64:631-8.
207. Li Q, Dashwood RH. Activator protein 2alpha associates with adenomatous polyposis coli/beta-catenin and Inhibits beta-catenin/T-cell factor transcriptional activity in colorectal cancer cells. J Biol Chem 2004;279:45669-75.
208. McPherson LA, Loktev AV, Weigel RJ. Tumor suppressor activity of AP2alpha mediated through a direct interaction with p53. J Biol Chem 2002;277:45028-33.

209. Li H, Watts GS, Oshiro MM, Futscher BW, Domann FE. AP-2alpha and AP-2gamma are transcriptional targets of p53 in human breast carcinoma cells. Oncogene 2006;25:5405-15.
210. Hensch T, Wargelius HL, Herold U, Strobel A, Oreland L, Brocke B. Electrophysiological and behavioral correlates of polymorphisms in the transcription factor AP-2beta coding gene. Neurosci Lett 2008;436:67-71.
211. Damberg M, Berggard C, Mattila-Evenden M, et al. Transcription factor AP-2beta genotype associated with anxiety-related personality traits in women. A replication study. Neuropsychobiology 2003;48:169-75.
212. Zhao F, Weismann CG, Satoda M, et al. Novel TFAP2B mutations that cause Char syndrome provide a genotype-phenotype correlation. Am J Hum Genet 2001;69:695-703.
213. Maeda S, Tsukada S, Kanazawa A, et al. Genetic variations in the gene encoding TFAP2B are associated with type 2 diabetes mellitus. J Hum Genet 2005;50:283-92.
214. Tsukada S, Tanaka Y, Maegawa H, Kashiwagi A, Kawamori R, Maeda S. Intronic polymorphisms within TFAP2B regulate transcriptional activity and affect adipocytokine gene expression in differentiated adipocytes. Mol Endocrinol 2006;20:1104-11.
215. Orso F, Cottone E, Hasleton MD, et al. Activator protein-2gamma (AP-2gamma) expression is specifically induced by oestrogens through binding of the oestrogen receptor to a canonical element within the 5'-untranslated region. Biochem J 2004;377:429-38.
216. Woodfield GW, Horan AD, Chen Y, Weigel RJ. TFAP2C controls hormone response in breast cancer cells through multiple pathways of estrogen signaling. Cancer Res 2007;67:8439-43.
217. Li H, Goswami PC, Domann FE. AP-2gamma induces p21 expression, arrests cell cycle, and inhibits the tumor growth of human carcinoma cells. Neoplasia 2006;8:568-77.
218. Guler G, Iliopoulos D, Guler N, Himmetoglu C, Hayran M, Huebner K. Wwox and Ap2gamma expression levels predict tamoxifen response. Clin Cancer Res 2007;13:6115-21.
219. Werling U, Schorle H. Transcription factor gene AP-2 gamma essential for early murine development. Mol Cell Biol 2002;22:3149-56.
220. Dalgin GS, Drever M, Williams T, King T, DeLisi C, Liou LS. Identification of novel epigenetic markers for clear cell renal cell carcinoma. J Urol 2008;180:1126-30.
221. Pike BL, Greiner TC, Wang X, et al. DNA methylation profiles in diffuse large B-cell lymphoma and their relationship to gene expression status. Leukemia 2008;22:1035-43.
222. Woodfield GW, Hitchler MJ, Chen Y, Domann FE, Weigel RJ. Interaction of TFAP2C with the estrogen receptor-alpha promoter is controlled by chromatin structure. Clin Cancer Res 2009;15:3672-9.
223. Chung W, Kwabi-Addo B, Ittmann M, et al. Identification of novel tumor markers in prostate, colon and breast cancer by unbiased methylation profiling. PLoS One 2008;3:e2079.
224. Cheng C, Ying K, Xu M, et al. Cloning and characterization of a novel human transcription factor AP-2 beta like gene (TFAP2BL1). Int J Biochem Cell Biol 2002;34:78-86.
225. Maehata T, Taniguchi H, Yamamoto H, et al. Transcriptional silencing of Dickkopf gene family by CpG island hypermethylation in human gastrointestinal cancer. World J Gastroenterol 2008;14:2702-14.
226. You J, Nguyen AV, Albers CG, Lin F, Holcombe RF. Wnt pathway-related gene expression in inflammatory bowel disease. Dig Dis Sci 2008;53:1013-9.
227. Kleibl Z, Fidlerova J, Kleiblova P, et al. Influence of dihydropyrimidine dehydrogenase gene (DPYD) coding sequence variants on the development of fluoropyrimidine-related toxicity in patients with high-grade toxicity and patients with excellent tolerance of fluoropyrimidine-based chemotherapy. Neoplasma 2009;56:303-16.
228. Sharma R, Hoskins JM, Rivory LP, et al. Thymidylate synthase and methylenetetrahydrofolate reductase gene polymorphisms and toxicity to capecitabine in advanced colorectal cancer patients. Clin Cancer Res 2008;14:817-25.

References

229. Tsuruta M, Nishibori H, Hasegawa H, et al. Heat shock protein 27, a novel regulator of 5-fluorouracil resistance in colon cancer. Oncol Rep 2008;20:1165-72.
230. Grivicich I, Regner A, Zanoni C, et al. Hsp70 response to 5-fluorouracil treatment in human colon cancer cell lines. Int J Colorectal Dis 2007;22:1201-8.
231. Ravizza R, Molteni R, Gariboldi MB, Marras E, Perletti G, Monti E. Effect of HIF-1 modulation on the response of two- and three-dimensional cultures of human colon cancer cells to 5-fluorouracil. Eur J Cancer 2009;45:890-8.
232. Chen YY, Wang ZX, Chang PA, et al. Knockdown of focal adhesion kinase reverses colon carcinoma multicellular resistance. Cancer Sci 2009.
233. Yang L, Hamilton SR, Sood A, et al. The previously undescribed ZKSCAN3 (ZNF306) is a novel "driver" of colorectal cancer progression. Cancer Res 2008;68:4321-30.
234. Karasawa H, Miura K, Fujibuchi W, et al. Down-regulation of cIAP2 enhances 5-FU sensitivity through the apoptotic pathway in human colon cancer cells. Cancer Sci 2009;100:903-13.
235. Shin YK, Yoo BC, Hong YS, et al. Upregulation of glycolytic enzymes in proteins secreted from human colon cancer cells with 5-fluorouracil resistance. Electrophoresis 2009;30:2182-92.
236. Akhdar H, Loyer P, Rauch C, Corlu A, Guillouzo A, Morel F. Involvement of Nrf2 activation in resistance to 5-fluorouracil in human colon cancer HT-29 cells. Eur J Cancer 2009.
237. Xi Y, Formentini A, Nakajima G, Kornmann M, Ju J. Validation of biomarkers associated with 5-fluorouracil and thymidylate synthase in colorectal cancer. Oncol Rep 2008;19:257-62.
238. Tanaka S, Sakai A, Kimura K, et al. Proteomic analysis of the basic proteins in 5-fluorouracil resistance of human colon cancer cell line using the radical-free and highly reducing method of two-dimensional polyacrylamide gel electrophoresis. Int J Oncol 2008;33:361-70.
239. Fazzone W, Wilson PM, Labonte MJ, Lenz HJ, Ladner RD. Histone deacetylase inhibitors suppress thymidylate synthase gene expression and synergize with the fluoropyrimidines in colon cancer cells. Int J Cancer 2009;125:463-73.
240. Niv Y. Microsatellite instability and MLH1 promoter hypermethylation in colorectal cancer. World J Gastroenterol 2007;13:1767-9.
241. Humeniuk R, Mishra PJ, Bertino JR, Banerjee D. Epigenetic reversal of acquired resistance to 5-fluorouracil treatment. Mol Cancer Ther 2009.
242. Morita S, Iida S, Kato K, Takagi Y, Uetake H, Sugihara K. The synergistic effect of 5-aza-2'-deoxycytidine and 5-fluorouracil on drug-resistant tumors. Oncology 2006;71:437-45.
243. Murakami J, Lee YJ, Kokeguchi S, et al. Depletion of O6-methylguanine-DNA methyltransferase by O6-benzylguanine enhances 5-FU cytotoxicity in colon and oral cancer cell lines. Oncol Rep 2007;17:1461-7.
244. Chung SK, Lee MG, Ryu BK, et al. Frequent alteration of XAF1 in human colorectal cancers: implication for tumor cell resistance to apoptotic stresses. Gastroenterology 2007;132:2459-77.
245. Lerebours F, Olschwang S, Thuille B, et al. Fine deletion mapping of chromosome 8p in non-small-cell lung carcinoma. Int J Cancer 1999;81:854-8.
246. Bova GS, MacGrogan D, Levy A, Pin SS, Bookstein R, Isaacs WB. Physical mapping of chromosome 8p22 markers and their homozygous deletion in a metastatic prostate cancer. Genomics 1996;35:46-54.
247. Arbieva ZH, Banerjee K, Kim SY, et al. High-resolution physical map and transcript identification of a prostate cancer deletion interval on 8p22. Genome Res 2000;10:244-57.
248. Cooke SL, Pole JC, Chin SF, Ellis IO, Caldas C, Edwards PA. High-resolution array CGH clarifies events occurring on 8p in carcinogenesis. BMC Cancer 2008;8:288.

References

249. MacGrogan D, Levy A, Bova GS, Isaacs WB, Bookstein R. Structure and methylation-associated silencing of a gene within a homozygously deleted region of human chromosome band 8p22. Genomics 1996;35:55-65.
250. Li Q, Jedlicka A, Ahuja N, et al. Concordant methylation of the ER and N33 genes in glioblastoma multiforme. Oncogene 1998;16:3197-202.
251. Bookstein R, Bova GS, MacGrogan D, Levy A, Isaacs WB. Tumour-suppressor genes in prostatic oncogenesis: a positional approach. Br J Urol 1997;79 Suppl 1:28-36.
252. Pils D, Horak P, Gleiss A, et al. Five genes from chromosomal band 8p22 are significantly down-regulated in ovarian carcinoma: N33 and EFA6R have a potential impact on overall survival. Cancer 2005;104:2417-29.
253. Guervos MA, Marcos CA, Hermsen M, Nuno AS, Suarez C, Llorente JL. Deletions of N33, STK11 and TP53 are involved in the development of lymph node metastasis in larynx and pharynx carcinomas. Cell Oncol 2007;29:327-34.
254. De Troyer A. Differential control of the inspiratory intercostal muscles during airway occlusion in the dog. J Physiol 1991;439:73-88.
255. Garshasbi M, Hadavi V, Habibi H, et al. A defect in the TUSC3 gene is associated with autosomal recessive mental retardation. Am J Hum Genet 2008;82:1158-64.
256. Alexander J, Bey E, Whitcutt JM, Gear JH. Adaptation of cells derived from human malignant tumours to growth in vitro. S Afr J Med Sci 1976;41:89-98.
257. Zhou H, Clapham DE. Mammalian MagT1 and TUSC3 are required for cellular magnesium uptake and vertebrate embryonic development. Proc Natl Acad Sci U S A 2009;106:15750-5.
258. Yuen RK, Avila L, Penaherrera MS, et al. Human placental-specific epipolymorphism and its association with adverse pregnancy outcomes. PLoS One 2009;4:e7389.
259. Shen L, Toyota M, Kondo Y, et al. Integrated genetic and epigenetic analysis identifies three different subclasses of colon cancer. Proc Natl Acad Sci U S A 2007;104:18654-9.
260. Ahuja N, Li Q, Mohan AL, Baylin SB, Issa JP. Aging and DNA methylation in colorectal mucosa and cancer. Cancer Res 1998;58:5489-94.
261. Belshaw NJ, Elliott GO, Foxall RJ, et al. Profiling CpG island field methylation in both morphologically normal and neoplastic human colonic mucosa. Br J Cancer 2008;99:136-42.
262. Suehiro Y, Wong CW, Chirieac LR, et al. Epigenetic-genetic interactions in the APC/WNT, RAS/RAF, and P53 pathways in colorectal carcinoma. Clin Cancer Res 2008;14:2560-9.
263. Yang E, Kang HJ, Koh KH, Rhee H, Kim NK, Kim H. Frequent inactivation of SPARC by promoter hypermethylation in colon cancers. Int J Cancer 2007;121:567-75.
264. Bradshaw AD, Graves DC, Motamed K, Sage EH. SPARC-null mice exhibit increased adiposity without significant differences in overall body weight. Proc Natl Acad Sci U S A 2003;100:6045-50.
265. Wang CS, Lin KH, Chen SL, Chan YF, Hsueh S. Overexpression of SPARC gene in human gastric carcinoma and its clinic-pathologic significance. Br J Cancer 2004;91:1924-30.
266. Franke K, Carl-McGrath S, Rohl FW, et al. Differential Expression of SPARC in Intestinal-type Gastric Cancer Correlates with Tumor Progression and Nodal Spread. Transl Oncol 2009;2:310-20.
267. Croonquist PA, Linden MA, Zhao F, Van Ness BG. Gene profiling of a myeloma cell line reveals similarities and unique signatures among IL-6 response, N-ras-activating mutations, and coculture with bone marrow stromal cells. Blood 2003;102:2581-92.
268. Hohenester E, Sasaki T, Giudici C, Farndale RW, Bachinger HP. Structural basis of sequence-specific collagen recognition by SPARC. Proc Natl Acad Sci U S A 2008;105:18273-7.
269. Volmer MW, Radacz Y, Hahn SA, et al. Tumor suppressor Smad4 mediates downregulation of the anti-adhesive invasion-promoting matricellular protein SPARC: Landscaping activity of Smad4 as revealed by a "secretome" analysis. Proteomics 2004;4:1324-34.

270. Koblinski JE, Kaplan-Singer BR, VanOsdol SJ, et al. Endogenous osteonectin/SPARC/BM-40 expression inhibits MDA-MB-231 breast cancer cell metastasis. Cancer Res 2005;65:7370-7.
271. Tai IT, Dai M, Owen DA, Chen LB. Genome-wide expression analysis of therapy-resistant tumors reveals SPARC as a novel target for cancer therapy. J Clin Invest 2005;115:1492-502.
272. Cheetham S, Tang MJ, Mesak F, Kennecke H, Owen D, Tai IT. SPARC promoter hypermethylation in colorectal cancers can be reversed by 5-Aza-2'deoxycytidine to increase SPARC expression and improve therapy response. Br J Cancer 2008;98:1810-9.
273. Ma S, Sang Q, Lanciego JL, Gundlach AL. Localization of relaxin-3 in brain of Macaca fascicularis: identification of a nucleus incertus in primate. J Comp Neurol 2009;517:856-72.
274. van der Westhuizen ET, Werry TD, Sexton PM, Summers RJ. The relaxin family peptide receptor 3 activates extracellular signal-regulated kinase 1/2 through a protein kinase C-dependent mechanism. Mol Pharmacol 2007;71:1618-29.
275. Tanaka M. Relaxin-3/insulin-like peptide 7, a neuropeptide involved in the stress response and food intake. Febs J 2010;277:4990-7.
276. Gundlach AL, Ma S, Sang Q, et al. Relaxin family peptides and receptors in mammalian brain. Ann N Y Acad Sci 2009;1160:226-35.
277. Smith CM, Shen PJ, Banerjee A, et al. Distribution of relaxin-3 and RXFP3 within arousal, stress, affective, and cognitive circuits of mouse brain. J Comp Neurol 2010;518:4016-45.
278. Huang YW, Luo J, Weng YI, et al. Promoter hypermethylation of CIDEA, HAAO and RXFP3 associated with microsatellite instability in endometrial carcinomas. Gynecol Oncol 2010;117:239-47.

7. Tables, Figures, Abbreviations

Tables

Table 1 Colorectal cancer classification systems 9
Table 2 Survival Rates: Colorectal Cancer (according to the National Cancer Institute) 12
Table 3 DNMT inhibitors 34
Table 4 Inhibitory profile of HDAC inhibitors 36
Table 5 The following chart gives an overview for all candidate marker genes 65
Table 6 Literature/Functional Classification Overview (as at December 2010) 66
Table 7 The following list ilustrates the results of the cell line screening (3 best markers) ... 67
Table 8 Cell line screening continued (all other markers) 69
Table 9 ROC results 70
Table 10 Correlation of TFAP2E and DKK4 expression and methylation 71
Table 11 Results of the mutation analysis for RXFP3, TUSC3 and TFAP2E 76
Table 12 Overview over the validation of downstream targets for TFAP2E 78
Table 13 Overview over the validation of downstream targets for TUSC3 80
Table 14 Overview over the validation of downstream targets for RXFP3 80
Table 15 Validation results for TFAP2E in 4 independent patient cohorts 89
Tables S1a-c Primer sequences for TFAP2E, TUSC3, RXFP3 and related genes 122

Figures

Figure 1 Intestinal anatomy - picture courtesy National Cancer Institute 3
Figure 2 The colorectal adenoma–carcinoma sequence 4
Figure 3 Graphical overview of colorectal cancer staging 9
Figure 4 Interplay between DNA methylation, histone covalent modifications 15
Figure 5 Histone modifications and their influence on chromation formation 17
Figure 6 Methylation of cytosine in DNA occurs at CG dinucleotides 20
Figure 7 DNA methylation patterns 23
Figure 8 Changes in DNA Methylation during the transformation into cancer cells 28
Figure 9 The classical model of cancer versus the epigenetic progenitor model of cancer 29
Figure 10 Histone modifications in cancer 30
Figure 11 Epigenetically acting drugs 31
Figure 12 Possible applications of DNA hypermethylation markers 41
Figure 13 Overview over bisulfite conversion and subsequent amplification 46
Figure 14 Principle of relative quantification by real time PCR 47
Figure 15 Overview over HRM and MethyLight technology 49
Figure 16 Examplary flourescent data from the melting standards for the TFAP2E HRM ... 50
Figure 17 Principle of reverse transcription 52
Figure 18 Expression and Methylation of CRC cell lines for TFAP2E and DKK4 68
Figure 19 Expression and Methylation of CRC cell lines for TUSC3 and SPARC 68
Figure 20 Expression and Methylation of CRC cell lines for RXFP3 68
Figure 21 ROC Curves for all there selected markers 70
Figure 22 Methylation results of 74 patients for TFAP2E, TUSC3 and RXFP3 72
Figure 23 Statistical analyses for correlation with clinical characteristics for RXFP3 73
Figure 24 Statistical analyses for correlation with clinical characteristics for TFAP2E 74
Figure 25 Statistical analyses for correlation with clinical characteristics for TUSC3 75
Figure 26 Significant correlations of APC and TP53 mutations for TUSC3 and TFAP2E 76

Figure 27 Dose response curves for determination of G418 concentration. 77
Figure 28 Validiation of genes with a more than 3fold change in microarray via qPCR. 79
Figure 29 Proliferation of stable overexpressing TFAP2E SW480 cell clones 81
Figure 30 Apoptosis of stable Apoptosis overexpressing TFAP2E SW480 cell clones 81
Figure 31 Dose response curves for SW480 cells and 5-FU concentration. 82
Figure 32 TFAP2E SW480 cells transiently transfected and treated with oxaliplatin. 82
Figure 33 TFAP2E SW480 cells transiently transfected and treated with irinotecan. 83
Figure 34 Resistance to treatment with fluorouracil (5-FU) irinotecan and oxaliplatin 83
Figure 35 Luciferase reporter assays 84
Figure 36 Chromatin immunoprecipitation 85
Figure 37 Migration and Adhesion Assays 85
Figure 38 Adhesion of TUSC3 clones on a collagen Type I coated plastic surface 86
Figure 39 Percent of migrating cells through a matrigel chamber 86
Figure 40 TUSC3 – Proliferation measured via MTT assays 87
Figure 41 TUSC3 – Apoptosis via TNF treatment and MTT assays. 87
Figure 42 Proliferation of RXFP3 SW480 cell clones measured via BRDU assays. 88
Figure 43 Meta-analyses on 5-FU response and TFAP2E methylation. 89
Figure 44 ROC curve to define the optimal cutoff for TFAP2E methylation 90
Figure 45 ROC curve on TFAP2E methylation between responder and non-responder 90

Tables, Figures and Abbreviations Lists

Abbreviations

5-AZA	5-Azacytidine
ABCG2	ATP-binding cassette, sub-family G, member 2
ACTB	Actin, beta
AID	Activation-induced cytidine deaminases
AJCC	American Joint Committee on Cancer
ALU	Short interspersed repetitive elements
AML	Acute myeloid leukemia
APC	Adenomatosis polyposis coli
APOBEC	Apolipoprotein B mRNA editing enzyme
AP-PCR	Arbitrarily primed PCR
ATCC	American Type Culture Collection
AUG	Translation start codon of a gene
BCNU	Bis-chloronitrosourea (carmustine)
BHQ1	Blackhole Quencher 1
BIRC3	Baculoviral IAP repeat containing 3
BRAF	v-raf murine sarcoma viral oncogene homolog B1
BrdU	Bromodeoxyuridine
BSA	Bovine serum albumin
CACO-2	Cell line, human colon adenocarcinoma
cAMP	Cyclic adenosine monophosphate
C5	Carbon 5
CACNA1G	calcium channel, voltage-dependent, T type, alpha 1G
CDH1	E-cadherin (epithelial)
CDKN2A/B	Cyclin-dependent kinase inhibitor 2A/B
cDNA	Complementary desoxyribonucleic acid
CDS	Coding sequence of a gene
CEA	Carcinoembryonic antigen
CH3	Methyl group
ChIP	Chromatin Immunoprecipitation
CIMP	CpG island methylator phenotype
CIN	Chromosomal instability
CML	Chronic myeloid leukemia

CO2	Carbon dioxide
COBRA	Combined bisulfite restriction assay
COL6A3	Collagen, type VI, alpha 3
COX2	Cyclooxygenase-2
CpG	Cytosine-guanine dinucleotide
CRC	Colorectal Cancer
CREBBP	CREB binding protein
C-terminus	Carboxy-terminus
CT	X-ray Computed Tomography
CTNNB1	catenin, beta 1, beta-catenin
CY5, CY3	Cyanine Dyes
DAPI	4',6-diamidino-2-phenylindole fluorescent stain
DAPK	Death-associated protein kinase
DGGE	Denaturing gradient gel electrophoresis
DH5 alpha	Escherichia coli strain
DHPLC	Denaturing High Performance Liquid Chromatography
DICER	Dicer 1, ribonuclease type III complex
DMEM	Dulbeccos modified Eagle medium
DMH	Differential Methylation Hybridization
DMSO	Dimethyl sulfoxid
DNA	Desoxyribonucleic acid
dsDNA	Double stranded DNA
DKK4	Dickkopf homolog 4 (Xenopus laevis)
DLD-1	Cell line, human colon adenocarcinoma
DMH	Differential Methylation Hybridization
DNMT	DNA-methyltransferase
dNTP	Desoxyribonucleosidetriphosphate
DPYP	Dihydropyrimidine dehydrogenase
DROSHA	Drosha, ribonuclease type III
DZNep	3-deazaneplanocin
ECM	Extracellular matrix
EGCG	Epigallocatechin gallate
EDTA	Ethylenediaminetetraacetic acid

EGFR	Epidermal growth factor receptor
EIF4E	Eukaryotic translation initiation factor 4E
EHMT2	G9a histone methyltransferase
ELISA	Enzyme-linked immunosorbent assay
EMP3	Epithelial membrane protein 3
EP300	E1A binding protein p300
ERBB2	v-erbb2 erythroblastic leukemia viral oncogene 2
ERK	Extracellular signal-regulated kinases
ERMA	Enzymatic regional methylation assay
ES	Embryonic stem
ESR1	Estrogen receptor 1 (also known as ER alpha)
EZH2	Enhancer of zeste homolog 2
FAM	6-carboxyfluorescein
FAP	Familial adenomatous polyposis
FBS	Foetal bovine serum
FDA	Food and Drug Administration
FIT	immunochemical Fecal Occult Blood Test
FLAG(-tag)	FLAG octapeptide, a polypeptide protein tag
FOBT	Fecal Occult Blood Test, gFOBT (guiac based)
FOLFOX	Folinic acid, Fluorouracil, Oxaliplatin (chemotherapy)
FOLFIRI	Fluorouracil, Folinic acid, Irinotecan
fw	Forward
FFPE	Formalin fixed paraffin embedded
G418	Geneticin, a aminoglycoside antibiotic
gDNA	Genomic DNA
GATA4	GATA binding protein 4
GKM	Gate-keeper mutation
GSTP1	Glutathione S-transferase pi 1 gene
H2A	Histone-2
H2B	Histone-2B
H3	Histone-3
H3K4	Histone-3 lysine-4
H3K9	Histone-3 lysine-9

H4	Histone-4
H4K20	Histone-4 lysine-20
HAT	Histone acetyltransferase
HCL	Hydrochloric acid
HCT-116	Cell line, human colon carcinoma
HDAC	Histone deacetylase
HEK	Human embryonic kidney cells
HIC1	Hypermethylated in cancer 1 gene
HIF1A	Hypoxia inducible factor 1, alpha subunit
HMT	Histone methyltransferase
HNPCC	hereditary nonpolyposis colorectal cancer
HOX	Homeobox genes
HpaII	HpaII Methyltransferase, recognizes CCGG
HPCE	High Performance Capillary Electrophoresis
HPLC	High Performance Liquid Chromatography
HRM	High resolution melting
HSP27	Heat shock protein 27
HSP70	Heat shock protein 70
HT-29	Cell line, human colon adenocarcinoma
IGF2	Insulin-like growth factor 2 (somatomedin A)
IPEM	Incomplete primer extension mixture
kDa	Dalton, (symbol: Da) unified atomic mass unit
KDM1A	Lysine (K)-specific demethylase 1A
KRAS	Kirsten rat sarcoma-2 viral oncogene homologue
LBH589	HDAC inihibitor, a hydroxamic acid
LC480	LightCycler 480
LINE	Long interspersed elements
LOH	Loss of heterozygosity
LOI	Loss of Imprinting
LOVO	A human colon adenocarcinoma cell line
LRP5/6	Low density lipoprotein receptor-related proteins
M	Methylated
MAC	Modified Astler-Coller classification

MAPK	Mitogen-activated protein kinases, also ERKs
MBD	Methyl-CpG binding domain protein
MCA	Methylated CpG island amplification
mRNA	Messenger ribonucleic acid
miRNA	microRNA
miRISC	RNA-induced silencing complex
MGCD0103	A benzamide histone deacetylase inhibitor
MDS	Myelodysplastic syndromes
MG98	Antisense inhibitor of human DNMT1
MGMT	O-6-methylguanine-DNA methyltransferase
MLH1	human mutL homolog 1 (also hMLH1) gene
MLL	myeloid/lymphoid or mixed-lineage leukemia (trithorax homolog, Drosophila)
MMR	Mis-Match Repair
MMP7	Matrix metallopeptidase 7 (matrilysin, uterine)
MRI	Magnetic resonance imaging
MS-275	A benzamide histone deacetylase inhibitor
MSI	Microsatellite instability
MS-ISH	Methylation-specific in situ hybridization
MSP	Methylation-specific PCR
MspI	MspI Methyltransferase, recognizes CCGG
MS-HRM	Methylation specific high resolution melting
MS-REs	Methylation-sensitive restriction endonuclease
MS-SnuPE	Methylation-sensitive single nucleotide primer extension
MS-SSCP	Methylation-sensitive single-strand conformational polymorphism analysis
MTHFR	Methylenetetrahydrofolate reductase (NAD(P)H)
MTT	3-(4,5-Dimethylthiazol-2-yl)-2,5-diphenyltetrazolium bromide
MTX	Methotrexate
MUTYH	mutY homolog (E. coli), A/G-specific adenine DNA glycosylase, a mismatch repair gene

MYST	MYST histone acetyltransferase
N	Nitrogen
NAD	Nicotinamide adenine dinucleotide
NCBI	National Center for Biotechnology Information
NDRG4	NDRG family member 4
NEUROG1	Neurogenin 1
NFE2L2	Nuclear factor (erythroid-derived 2)-like 2
NS-ARMR	Non nonsyndromic autosomal recessive mental retardation
NT	Non-tumor
N-terminal	Amino-terminal
NUPR1	Nuclear protein, transcriptional regulator, 1
OH	Hydroxy
ONC	Oncogene
ORC6	Origin recognition complex, subunit 6
OST	N-oligosaccharyltransferase
OSMR	Oncostatin M receptor
PBL	Peripheral blood lymphocytes
PBS	Phosphate buffered saline
PCR	Polymerase chain reaction
PET	Positron emission tomography
PI3K	Phosphoinositide 3-kinase
PKC	Protein kinase C
PKM2	Pyruvate kinase, muscle
PMR	Percentage of methylated reference
POLB	Polymerase (DNA directed), beta
PSAT1	Phosphoserine aminotransferase 1
PTK2	Protein tyrosine kinase 2
PXD101	HDAC inhibitor, a hydroxamic acid
RARB	Retinoic acid receptor beta
RECIST	Response Evaluation Criteria In Solid Tumors
RFA	Radio-frequency ablation
RG108	N-Phthalyl-L-tryptophan

Tables, Figures and Abbreviations Lists

RLU	Relative light unit
RNA pol	RNA polymerases
ROC	Receiver operating characteristic curve
RPD3	Rpd3 histone deacetylase (yeast)
RRM2	Ribonucleotide reductase M2
RsaI	Restriction Endonuclease, recognizes GTAC
RT-PCR	Reverse transcription polymerase chain reaction
RUNX3	Runt-related transcription factor 3
RXFP3	Relaxin family peptide 3
SAHA	Suberoylanilide hydroxamic acid
SALPR	Somatostatin and Angiotensin-Like Peptide Receptor
SAM	S-adenosylmethionine
SDS	Sodium lauryl sulfate, an anionic surfactant
SDS-PAGE	Sodium dodecyl sulfate polyacrylamide gel electrophoresis
SERTAD1	SERTA domain containing 1 gene
SFRP1	Secreted frizzled-related protein 1
SFRP2	Secreted frizzled-related protein 2
SIRT1-7	Sirtuins, NAD-dependent deacetylases
SGI-1027	Small molecule inhibitor, quinoline-based
SLC19A1	Ssolute carrier family 19 (folate transporter), member 1
SmaI	Restriction Endonuclease, recognises CCCGGG
SMYD3	SET and MYND domain containing 3
SOCS1	Suppressor of cytokine signaling 1
SssI	Methyltransferase gene from *Spiroplasma* sp. strain MQ1
STOP codon	Termination codon within mRNA that signals a termination of translation
SYBR	SG cyanine dye
SW480	Cell line, human colon adenocarcinoma
TAE	Tris-acetate
TDG	Thymine-DNA glycosylase
TESS	Transcription Element Search System

TET1	Tet oncogene 1
TF	Transcription Factor
TFAP2E	Transcription factor activating enhancer binding protein 2 epsilon
TFPI2	Tissue factor pathway inhibitor 2
TGF beta	Transforming growth factor, beta
TNF alpha	Tumor necrosis factor alpha
TNM	Tumors/Nodes/Metastases cancer staging system
TOP10	Escherichia coli strain
TP53	Tumor protein 53
TPG	Tumor progenitor gene
TRDMT1	tRNA aspartic acid methyltransferase 1
TSA	Trichostatin A
TSG	Tumor suppressor gene
TSS	Transcription start site of a gene
TU	Tumor
TUSC3	Tumor suppressor candidate 3
TYMP	Thymidine phosphorylase
TYMS	Thymidylate synthetase
U	Unmethylated
UCK2	Uridine-cytidine kinase 2
UICC	Union for International Cancer Control
UMPS	Uridine monophosphate synthetase
UPP1	Uridine phosphorylase 1
USA	United States of America
VEGF	Vascular endothelial growth factor A
VIM	Vimentin
VHL	Von Hippel-Lindau tumor suppressor gene
WNT	Wingless-type MMTV integration site family
WRN	Werner syndrome, RecQ helicase-like
XAF1	XIAP associated factor 1
XmaI	Restriction Endonuclease, isoschizomer of SmaI
ZKSCAN3	Zinc finger with KRAB and SCAN domains 3

Appendix A

8. Appendices

Table S1a Primer sequences for TFAP2E, TUSC3, RXFP3 and related genes.

Gene/ orientation	Primer sequence (5'-3')	Product length	Accession/ Location	Usage
DKK4			Genbank	
Forward	ATATTAGAAAGGCAGCTTGATGAG	206bp	NM_014420.2	Expression Verification
Reverse	TTACAAATTTTCGTCCAAAAATGAC	bp 406-611	Exon 3-4	
Forward	GAAAGGGATGAAGCAGAAGTTTTA	1kb	NC_000008.10	Luciferase
Reverse	GTCGTCTGTTTGTCACTGCTTTT	2 put. AP2 BS	Promoter	
Forward	CTCCCAAAGTGCTGGGATTA	2kb	NC_000008.10	Luciferase
Reverse	GCACGTCGTCTGTTTGTCAC	4 put. AP2 BS	Promoter	
Forward	TTTAAGCGGTTGGGATTTTG	500bp	NC_000008.10	Luciferase
Reverse	TAACCAGATGTGCCTCCTCC	2 put. AP2 BS	Promoter	
Forward	TTCGCCTGTGTATATTGCCA	114bp	NC_000008.10	ChIP 1
Reverse	GATAAAGGAAAGAGCCCCCA	2 put. AP2 BS	Promoter	
Forward	TAAGCGGTTGGGATTTTGAC	300bp	NC_000008.10	ChIP 2
Reverse	GGCAGAGCAGGATGTCTGTA	2 put. AP2 BS	Promoter	
Forward	CGGGATCCGCCGCCACCATGGTGGCGGCCGTCCTGCT	674bp	NM_014420.2	Cloning CDS
Reverse	ATAAGAATGCGGCCGCTTATAGCTTTTCTATTTTTTGGCATAC	bp 112-786	Exon1-4	
TFAP2E				
Forward	TTTAGAAGCGGTTTTCGTATC	139bp	NC_000001.10	MethyLight/ MS-HRM
Reverse	CCGAACGCTTACCTACAATC	CpG Island	Intron3	
Probe	TTGCGGTGGGCGTTTTCGGGTT	3'BHQ1-5'FAM	+4266- +4405	
Forward	TAGACCAGTCCGTGATCAAGAAAGT	310bp	NM_178548.3	Expression
Reverse	AGGTTGAGCCCAATCTTCTCTAAC	bp 747-1056	Exon 3-5	
Forward	CACCTACTCCGCCATGGAG	1331bp	NM_178548.3	Cloning
Reverse	GTGGGAGAAGCAGTTATTCCG	CDS	Exon 1-7	
Forward	GTTTTGATTAATGTGGGTTGAATTTA	753bp	NC_000001.10	BSP
Reverse	CAACCTAAAAAAATCCTCCTCAAC	CpG Island	Promoter/Exon 1	
Reverse	TTTATCATCATCATCTTTATAGTCTTTCCG	FLAG Epitope	CDS	Cloning
Reverse	TTATTTATCATCATCATCTTTATAGTCTTTCCGATG	FLAG + STOP	CDS	
ACTB				
Forward	TGGTGATGGAGGAGGTTTAGTAAGT	132bp	NG_007992.1	MethyLight total gDNA input control
Reverse	AACCAATAAAACCTACTCCTCCCTTAA	no CpGs	Promoter	
Probe	ACCACCACCCAACACACAATAACAAACACA	3'TAMRA-5'FAM	−1599/−1467	
TFAP2A				
Forward	CTCGATCCACTCCTTACCTCAC	396bp	NM_003220.2	Expression control
Reverse	CCTGCAGGCAGATTTAATCCTA	bp 693-1088	Exon 4-7	
TFAP2B				
Forward	AATGGAAGACGTCCAGTCAGTT	431bp	NM_003221.3	Expression control
Reverse	AGTGAACAGCTTCTCCTTCCAC	bp 691-1121	Exon 2-6	
TFAP2C				
Forward	GATCAGACAGTCATTCGCAAAG	388bp	NM_003222.3	Expression control
Reverse	CAAAGTCCCTAGCCAAATGAAC	bp 808-1195	Exon 3-6	
TFAP2D				

Appendix A

Forward	GCCAAGGTGGAGTGATAAGAAG	392bp	NM_172238.3	Expression
Reverse	GGCAAGATGTTCTCCTACTGCT	bp 1087-1478	Exon 3-6	control
PCDH10				
Forward	TAAATAGGGGAATTTTTTTATTTTTTTT	355bp	NC_000004.11	HRM
Reverse	TCCTTCCTCCTACTTCAACCTCTAAAC	-321 to +34	Promoter/Exon 1	BSP
Forward	TAGTCGTTTTGGCGGCG	78bp	NM_032961.1	
Reverse	AACGCACGACCAACACGA	+2949 to +3027	NM_020815.1	MethyLight
Probe	AAAAAACACCGAACCCAACGCGATAATAAA		Exon 1	
Forward	CTTATGAGCTGGTGATCGAGGT	400bp	NM_032961.1	Expression
Reverse	TCTGACTTGCTGAGTTTCTTCTTG	bp 2784-3183	Exon 1	Isoform A
Forward	CAAGACCGACCTGATGTTTCTTA	275bp	NM_020815.1	Expression
Reverse	CTGCTCCCACAACGATACACATA	bp 3316-3590	Exon 1	Isoform B
KRT13				
Forward	GCTTTGTTGACTTTGGTGCTT	500bp	NM_002274.3	Expression
Reverse	CACCTGGTTGCTAAATTCCTTC	326-825 Exon 1-4	NM_153490.2	both isoforms
KRT14				
Forward	AGGAATGGTTCTTCACCAAGAC	420bp	NM_000526.4	Expression
Reverse	AGGTCACATCTCTGGATGACTG	bp 969-1388	Exon 4-8	control
KRT15				
Forward	CAGAATGCGACTACAGCCAATA	503bp	NM_002275.3	Expression
Reverse	CTGGATCATTTCTGTGTTGGAG	Exon 1-5	bp 507-1009	control
KRT16				
Forward	AACAGCGAACTGGTACAGAGC	363bp	NM_005557.3	Expression
Reverse	CTCGCGGGAAGAATAGGATT	Exon 5-7	bp 1103-1465	control
KRT17				
Forward	CTGCAGAACAAGATCCTCACAG	691bp	NM_000422.2	Expression
Reverse	GCAGGATTTTGTATTCCTGGTT	bp 469-1159	Exon 1-6	control
KRT18				
Forward	CAGATCTTCGCAAATACTGTGG	428bp Exon1-5/2-6	NM_000224.2	Expression
Reverse	ACTGTGGTGCTCTCCTCAATCT	bp 484-911	NM_199187.1	both variants
KRT19				
Forward	ACCATTGAGAACTCCAGGATTG	496 bp	NM_002276.4	Expression
Reverse	GCTCAATCTCAAGACCCTGAAG	bp 575-1070	Exon 2-5	control
KRT20				
Forward	CTAAACTGGCTGCTGAGGACTT	420bp	NM_019010.2	Expression
Reverse	CCGTTAGTTGAACCTCAGTTCC	bp 485-904	Exon 2-5	control

Table S1b Primer sequences (continued).

Gene/ orientation	Primer sequence (5'-3')	Product length	Accession/ Location	Usage
TUSC3				
Forward	CCGAACAAACGTAATACGCG	105bp	NG_012141.2	MethyLight/ MS-HRM
Reverse	ACGGCGTGAAGGAGCG	CpG Island	Exon 1	
Probe	TACGCGCGGTAGTCGTGCGC	3'BHQ1-5'FAM	bp +210 to 279	
Forward	GCAGCTGATGGAATGGAGTT	380bp	NM_006765.3	Expression
Reverse	ATCCGTTCTGTCAGCAATCC	Exon 2-4	NM_178234.2	both Isoforms
Forward	TAGATTGAGGTTTTAGGGTTAAAGGATTAT	450bp	NG_012141.2	BSP/HRM
Reverse	TACAAAACAACAACAACAAAAAAAA	CpG Island	Promoter/Exon 1	-76/+316TSS
Forward	AGGAGACACTGCCCTGCC	1100bp (324-1423)	NM_006765.3	Cloning

123

Appendix A

Reverse	TTTTTAAGTGCCATGGTCCAA	CDS isoform A	Exon 1-10	
Forward	AGACACTGCCCTGCCGCGAT	1075bp (327-1401)	NM_178234.2	Cloning
Reverse	ATCCCACTTGGCTTCATTTA	CDS isoform B	Exon 1-11	
Forward	AGCCAGGCCAGTTTGTGGC	transcript variant 1	NM_006765.2	Expression
Reverse	TGCCATGGTCCAAATCACATC	1027-1281	Exon 7-9	Isoform A
Forward	AGGATGGTTTTAGATTGAGGTTTTAGG	511bp	NG_012141.2	BSP
Reverse	CAAAAAAATCCATTCTACCTCCTTTTT	CpG Island	Promoter/Exon 1	-144 to +397
SPARC				
Forward	GGCGGAAAATCCCTGCCAGAACC	200bp	NM_003118.2	Expression
Reverse	CCTCCAGGGTGCACTTTGTGGC	bp 306-505	Exon 4-6	
Forward	AGGGTTCCCAGCACCATGAG	936bp	NM_003118	Cloning
Reverse	GGAGTGGATTTAGATCACAAGAT	bp 91-1026	CDS Exon 1-10	
Forward	AGGCACAGGAAAATCAGGTG	2100bp	NC_000005.9	Cloning
Reverse	GGCAACAGGAAACCACTCAG	-1943-157	Promoter	Luciferase
Forward	GTGCAGAGGAAACCGAAGAG	64bp	NM_003118.2	Expression
Reverse	TGTTTGCAGTGGTGGTTCTG	bp 268-341	Exon 4-5	realtime PCR
Forward	TTTCGCGGTTTTTTAGATTGTTC	135bp	NC_000005.9	
Reverse	AACGACGTAAACGAAAATATCG	+28 - 107		MethyLight
Probe	ACGACAAACAAAACGCGCTCTC	3'BHQ1-5'FAM	Exon 1/Intron 1	
Relaxin 3				
Forward	AATTCATCCGAGCAGTCATCTT	296bp	NM_080864.2	Expression
Reverse	ACTTTTGCTACACCCCCACTT	bp 113-408	Exon 1-2	
RXFP3				
Forward	ATTTCGGAAAGCGTTTTTCG	74bp	NC_000005.9	
Reverse	CAACTCCGAATAAATTACCAACGAC	CpG Island	Promoter	MethyLight
Probe	TTACGATAACACTTACACGACCAAAACGACGAA	3'BHQ1-5'FAM	bp -274 to -200	
Forward	AACTGGGGTAAACCGTGTTATCT	387bp	NM_016568.2	Expression
Reverse	GTTGGTGACGAAGAGGTTGATAG	bp 294-726	Exon 1	
Forward	TGTTGAAATTTTGGAGAGGAAAATTG	620bp	NC_000005.9	BSP
Reverse	AAAAACCCAAAAACTAAATACTAAAC	bp-385TSS +235	Prom./Exon 1	
Forward	GTGGGTTTTGTTTGTAGTTTAATTTTT	441bp	NC_000005.9	HRM
Reverse	CAAACTCGAAATCCCTAAATCCTTAT	bp -350 TSS +91	Prom. CpG Isle	
Forward	GTACCTGCGCATGCAGATG	1433bp	NM_016568.2	Cloning
Reverse	CCTGAGGCCTGCGTCAGTA	bp 358-1767	Exon 1 CDS	
Reverse	TTTATCATCATCATCTTTATAGTCGTAGGC	FLAG Epitope	CDS	Cloning
Reverse	TTATTTATCATCATCATCTTTATAGTCGTAGGCAGA	FLAG + STOP	CDS	
RXFP4				
Forward	CTATGCCAGCATCTTCCTCATC	453bp	NM_181885	Expression
Reverse	AGGTCAAACTTCACCAGGACAC	bp 396-848	Exon 1	
RXFP1				
Forward	TTCAGAAGCTGTACCTGCAAAA	443bp	NM_021634	Expression
Reverse	TTCATCCAGTTTCTGGAGAGGT	bp 537-979	Exon 5-11	

Table S1c Primer sequences (continued).

Gene/ orientation	Primer sequence (5'-3')	Product length	Accession/ Location	Usage
MUC13				
Forward	CTTGTTTAAAGATGTATTTGGCACA	261bp	NM_033049.2	MicroArray Verification
Reverse	GTCTGGTTACAGCCATAATAATCAC	bp 762-1022	Exons 5-7	
VIL1				

Appendix A

Forward	AAGTGGAGTAACACCAAATCCTATG	429bp	NM_007127.2	MicroArray
Reverse	AAAATTTCACTTCAATTGGTGTAGG	2217-2645	Exon 17-20	Verification
DDIT4				
Forward	ACTTGTCTTAGCAGTTCTCGC	376bp	NM_019058.2	MicroArray
Reverse	CACACAAGTGTTCATCCTCAG	bp 107-482	Exon 1-3	Verification
MAGEC2				
Forward	GGACCTCCCACCATAGAGAGAAGAA	202bp	NM_016249.2	MicroArray
Reverse	AGCAGCAGGTAAACGTATCAACAGG	bp 132-33	Exon 2-3	Verification
PARM1 (DKFZP564O0823)				
Forward	AAAATCAGGCATTCCTCCTATGGAA	201bp	NM_015393.2	MicroArray
Reverse	CAGGGCTTAAGATTGTCGTTCAATG	bp 1081-1281	Exon 2-3	Verification
ASB4				
Forward	AAGGCTATTTTGATCCAAAGGCAAA	422 bp	NM_016116.1	MicroArray
Reverse	TCTTGGTTGTTGGTCTTCATGTTCA	bp 100-521	Exon 1-2	Verification
ALD1H3				
Forward	AGTTTGCTACATGTAACCCTTCAAC	294 bp	NM_000693.2	MicroArray
Reverse	CTGCAAAGTATCTGAGGGTTCTAAT	bp 247-540	Exon 2-4	Verification
SERPINE1				
Forward	CTATGGGATTCAAGATTGATGACAA	264bp	NM_000602.2	MicroArray
Reverse	TGATCATACCTTTTGTGTGTGTCTT	bp 399-662	Exon 2-4	Verification
KITLG				
Forward	CTTGGATTCTCACTTGCATTTATCT	323 bp Exon 2-4	NM_000899.3	MicroArray
Reverse	CTCCACAAGGTCATCCACTATATTC	bp 200-522	NM_003994.4	Verification
ITGB8				
Forward	CCTCTGGGCAGCCTGGGTGT	320bp	NM_002214.2	MicroArray
Reverse	CTTCGGCTCCTGGACGCAGC	bp 783-1102	Exon 1-4	Verification
FAT1				
Forward	CGGCAGGTACCATGCGGACG	134bp	NM_005245.3	MicroArray
Reverse	CGGGAATCGCAAAGTTGGCCC	bp 44-177	Exon 1-2	Verification
UIMC1				
Forward	GCCTGAATAGTTGCCGGCCT	335bp	NM_016290.3	MicroArray
Reverse	TGTGTGCTCAGCCGAGTGGC	bp 517-851	Exon 1	Verification
NUPR1				
Forward	CCTCGGAGGTGGAGGCCGGA	296bp	NM_012385.2	MicroArray
Reverse	TAGCCCCTCAGAGACTCAGTCAGCG	375-670/429-724	NM_001042483.1	Verification
GEM				
Forward	GTGCGCTCAGCACTGGGATTTTCTG	746/736bpEx1-4/5	NM_181702.2	MicroArray
Reverse	GTCGCACAATGCCCTCAAACAGCTC	206-951/101-836	NM_005261.3	Verification
GPR35				
Forward	AGCTCTCCCAGGGCATCTACCTGAC	253bp	NM_005301.2	MicroArray
Reverse	GCCATGGAGTTGAAATTGTGCCGGG	bp 844-1096	Exon 1	Verification
CYP24A1				
Forward	GCCTATCGCGACTACCGCAAAGAAG	359bp	NM_001128915.1	MicroArray
Reverse	GTGACCATCATCCTCCCAAACGTGC	bp 804-1162	NM_000782.4	Verification
PSG5				
Forward	AAGGAGGAAGGACAGCACAGCCTAC	205bp	NM_002781.3	MicroArray
Reverse	CGTGACTTGAGCAGTGATAGGCAGG	bp 38-242	NM_001130014.1	Verification
PSG1				
Forward	CACAAGCAGCAGAGACCATGGGAAC	486bp	NM_006905.2	MicroArray

Appendix A

Reverse	TGGAGGGCTTAGGAGTCTCCAGGTG	bp 122-589	Exon 1-3	Verification
PSG7				
Forward	GCCCCTCCCTGCACACAGCATATAA	438bp	NM_002783.2	MicroArray
Reverse	GATGGAGGGTTTGGGAGTCTCCAGG	bp 118-555	Exon 1-3	Verification
Forward	CCCTCCATCTCCAGCAGCAATTTCA	563bp	NM_002783.2	Expression
Reverse	CTGGGGAGGTCTGGACCATAGAGGA	bp 547-1109	Exon 2-5	
TOMM22				
Forward	GCGGCCGGAGCCACTTTTGA	218bp	NM_020243.4	MicroArray
Reverse	GCCCCTGGCATTCCTCCTGAGA	bp 214-431	Exon 2-4	Verification
AP1M2				
Forward	GCCAACGGCAGCGTCCTTCT	266bp	NM_005498.4	MicroArray
Reverse	ACCTGGGTGCTGAGGCGGTA	bp 637-902	Exon 5-7	Verification
HNF4A				
Forward	GTCAGCGCCCTGTGTGCCAT	147bp	NM_178850.1	MicroArray
Reverse	CACCACGCACTGCCGGCTAA	bp 255-401	NM_000457.3	Verification
	NM_001030003.1+ NM_001030004.1	NM_175914.3	NM_178849.1	Isoforms1-6
NUP214				
Forward	GCCAGCGCAGGAGGATTCGG	222bp	Exon 33-36	MicroArray
Reverse	CCCTCAGCTTCGCCAGCCAC	bp 6166-6387	NM_005085.2	Verification
KIAA1199				
Forward	ATCCACATCTCAGAGGGAGGCAAGC	482bp	Exon 3-6	MicroArray
Reverse	GCAACAGAAAGGATCCTGCCATCGG	bp 483-964	NM_018689.1	Verification
SCNN1A				
Forward	CCCAGGAATGGGTCTTCCAGATGCT	293bp	NM_001159575.1	Verification
Reverse	TGGAGACCAGTATCGGCTTCGGAAC	bp 1934-2226	NM_001159576.1	NM_001038.5
CD24				
Forward	GCAGTCAACAGCCAGTCTCTTCGTG	142bp	NM_013230.2	MicroArray
Reverse	CTTGCCACATTGGACTTCCAGACGC	bp 296-437	Exon 1	Verification
ANKRD36B				
Forward	AGCCGACGATTAAGGAAGACGACGA	376bp	Exon 1-3	MicroArray
Reverse	GCGCCATTTTGCAGCAGAAGAGTTG	bp 234-609	NM_025190.3	Verification
HSPA6				
Forward	CCGCATTTCTTTCAGCAGCCTGAGT	249bp	NM_002155.3	MicroArray
Reverse	CACGGGCTTTTTATCCTTTTGCGCC	bp 31-279	Exon 1	Verification
CDH1				
Forward	CAACAAGCCCGAATTCACCCAGGAG	380bp	Exon 6-9	MicroArray
Reverse	CCTGACCCTTGTACGTGGTGGGATT	bp 895-1274	NM_004360.3	Verification
MUC21				
Forward	TTCCTCATCACCCTGGTCTCGGTTG	132bp	Exon 1-2	MicroArray
Reverse	ATGGTTGAGGCCATGAGGGTGGTAG	bp 1692-1823	NM_001010909.2	Verification
ZFAND2A				
Forward	TCCAGAAGGATGTTCACGTCCCAGT	222bp	NM_182491.2	MicroArray
Reverse	AGTTGCCGTGACATTGGGCACATAC	bp 444-665	Exon 3-5	Verification
ROCK1				
Forward	GTGACTGGTGGTCGGTTGGGGTATT	285bp	NM_005406.2	MicroArray
Reverse	GTGCTACAGTGTCTCGGAGCGTTTC	bp 1717-2001	Exon 7-10	Verification
LRRFIP				
Forward	TGAGAAGGGGTCTCGTAACATGCCG	272bp/344bp	NM_001137551.1	MicroArray

Appendix A

Reverse	CAGCCAGCTGTTCTTCAAGCTCCAG	bp 475-746	NM_001137550.1	Verification
	NM_001137553.1; NM_004735.3	Exon 13-16	NM_001137552.1	Isoforms 1-5
SCNN1A				
Forward	CAACAACGGTCTGTCCCTGATGCTG	202bp	NM_001159575.1	Verification
Reverse	ATAATCGCCCCCAAGTCTGTCCAGG	bp 1261-1462	NM_001159576.1	NM_001038.5
ANKRD37				
Forward	GGGTGCGGCGAGTGTCTCAC	150bp	NM_181726.2	MicroArray
Reverse	GATCCGGGGGTGCGTTGACC	bp 174-323	Exon 1-2	Verification
NRN1				
Forward	TCCTCGCGGTGCAAATAGCGTA	224bp	NM_016588.2	MicroArray
Reverse	CGCCCCTTCCTGGCAATCCG	bp 689-912	Exon 1-3	Verification
LPAR6 (P2RY5)				
Forward	TCCCTCTGCTATGGCTCTTCCTCAG	154bp	NM_005767.5	Verification
Reverse	GGCCTTTTCCTCAGTTGCCAGTTGT	827-980/345-474	NM_001162497.1	Exons 7-8b
Forward	GCACCGCAGAAGGCATTTCCACATA	269bp	NM_001162498.1	Isoform 1-2
Reverse	TGAGGCCTTTTCCTCAGTTGCCAGT	bp 184-428	Isoform 3	Exon 8a
GDF15				
Forward	CTGCTAACCAGGCTGCGGGC	143bp	NM_004864.2	MicroArray
Reverse	AGGGCGGCCCGAGAGATACG	bp 219-361	Exon 1-2	Verification
FN1				
Forward	CCTATGTGGTCGGAGAAACGTGGGA	400bp	NM_002026.2	Verification
Reverse	CCTTGTGTCTTCAGCCACTGCATCC	bp 847-1246	NM_212482.1	Exons4-7
	NM_212476.1; NM_212475.1; NM_212478.1	NM_212474.1	NM_054034.2	Isoforms1-7
EGLN3				
Forward	GGCTGCGAGGCCATCAGCTT	175bp	NM_022073.3	MicroArray
Reverse	AGCGACCATCACCGTTGGGG	bp 585-759	Exon 1-2	Verification
PDGFA				
Forward	GGCCGAGGAAGCCGAGATCC	156bp	NM_002607.5	MicroArray
Reverse	GTGGCATGGACCCCGTGAGC	bp 900-1055	NM_033023.4	Verification
SC4MOL				
Forward	TGGGCATGGGTGACCATTCGTT	107bp	NM_001017369.1	MicroArray
Reverse	TGCCGAGAACCAGCATAGAAAGGG	770-876/484-590	NM_006745.3	Verification
BNIP3L				
Forward	CGGCGGACTCGGCTTGTTGT	196bp	NM_004331.2	MicroArray
Reverse	GCTCCACCCAGGAACTGTTGAGG	bp 44-239	Exon 1-2	Verification
SNAI2				
Forward	GATGCCGCGCTCCTTCCTGG	218bp	NM_003068.4	MicroArray
Reverse	GGCCATTGGGTAGCTGGGCG	bp 164-381	Exon 1-2	Verification
DDIT3				
Forward	GGTGGCAGCGACAGAGCCAA	177-458bp	NM_001195055.1	MicroArray
Reverse	ACCAGGCTTCCAGCTCCCAG	bp 54- 230/511	NM_001195057.1	Verification
	NM_001195056.1; NM_001195053.1	NM_004083.5	NM_001195054.1	Variants 1-6
SERPINE2				
Forward	ACAACAGGGTCAGAAAACCTCCATGT	bp1225-1406/1409	185bp	MicroArray
Reverse	ACAGCACCTGTAGGATTATGTCGGATG	bp 1155-1336	Exons	Verification
	NM_001136529.1; NM_001136528.1	bp 1643-1824	NM_001136530.1	NM_006216.3
BIRC3				
Forward	TGCCAAGTGGTTTCCAAGGTGTGA	233bp	NM_001165.3	MicroArray

Appendix A

Reverse	AGCCCATTTCCACGGCAGCA	bp 3718-3931	Exon 3-6	Verification
TXNIP				
Forward	CCTCAGGGGCCTCTGGGAAC	230bp	NM_006472.3	MicroArray
Reverse	GCAGAGACAGACACCCGCCC	bp 657-886	Exon 2-4	Verification
NRIP1				
Forward	GGCGGCCTGGGGAAGTGTTT	208bp	NM_003489.3	MicroArray
Reverse	TCTCCAAGCTCTGAGCCTCTGC	bp 51-258	Exon 1-3	Verification
HBEGF				
Forward	CCGGGACCGGAAAGTCCGTG	244bp	NM_001945.2	MicroArray
Reverse	ACCCGGGTGGCAGATGCAGG	bp 443-686	Exon 2-4	Verification
7A5 (MACC1)				
Forward	AGGAGGGGTCACAGGTGAACGA	198bp	NM_182762.3	MicroArray
Reverse	CAAGTCTGGGTCCTGGCATTCTGT	bp 247-444	Exon 2-4	Verification
KCTD11				
Forward	CTGCCCCGTGGGTACGGAGA	135bp	NM_001002914.2	MicroArray
Reverse	GTGGAGCAGGGCAGCTGTGG	bp 1193-1327	Exon 1	Verification
ARRDC4				
Forward	CCGGCCGGTGAAGGCATCAT	143bp	NM_183376.2	MicroArray
Reverse	CGTTCCAACACTGCCCGCAC	bp 479-583	Exon 1-3	Verification
CARD11				
Forward	CTACAGCCGAGCCCAGCAGC	178bp	NM_032415.3	MicroArray
Reverse	GCTTGCTCGCGAGAGACGGG	bp 2906-3083	Exon 18-20	Verification
MOBKL2A				
Forward	AGACACGGTCCGCGGGAGAG	223bp	NM_130807.2	MicroArray
Reverse	TGTGCAGCTCGAAGCGCTGG	bp 193-415	Exon 2-3	Verification
EFCAB4B				
Forward	CGGAGCGGAAAGAGTCGGGC	161bp Exon1-2	NM_001144958.1	Verification
Reverse	CAGCCCAATGGCCCCAGGTG	bp 91-251	NM_001144959.1	NM_032680.3
CA9				
Forward	TATGGAGGCGACCCGCCCTG	225bp	NM_001216.2	MicroArray
Reverse	CCCGGGACCCAGAGCCATCT	bp 469-693	Exon 2-4	Verification
DKK1				
Forward	GATGGTAGCGGCGGCTCTCG	212bp	NM_012242.2	MicroArray
Reverse	CGCACGGGTACGGCTGGTAG	bp 199-410	Exon 1-2	Verification
ITGA10				
Forward	ATGTGCTGGATACATCAGATTACCT	624bp	NM_003637.3	Expression
Reverse	GCAGGGTAGACTCACTAGAGAACAG	bp 2301-2924	Exon 17-24	control
PPP1R15A				
Forward	GGCACTTGAGGCAGCCGGAG	180bp	NM_014330.3	MicroArray
Reverse	TGGCCTGGGGCCATGTGTCT	bp 104-283	Exon 1-2	Verification
PTK2 (FAK)				
Forward	GCCAACTTTGAATTTCTTCTATCAA	340bp	NM_001199649.1	Expression
Reverse	ATCAAATCTGTAGACTGGAGACAGG	bp 653-992	Exon 6-10	control
SMAD4				
Forward	CATAGTTTGATGTGCCATAGACAAG	429bp	Exon 3-6	Expression
Reverse	CACATATTCATCCTTCACCATCATA	bp 324-752	NM_005359.5	control
TFF1				
Forward	AGAATTGTGGTTTTCCTGGTGT	283bp	Exon 2-3	Expression

Appendix A

Reverse	CAGTCAATCTGTGTTGTGAGCC	bp 156-438	NM_003225.2	control
TFF2				
Forward	CAGTGTTTTGACAATGGATGCT	310bp	Exon 2-4	Expression
Reverse	TAAGGCGAAGTTTCTTCTTTGG	bp 314-581	NM_005423.4	control
TFF3				
Forward	CAAACAACGGTGCATAAATGAG	420bp	Exon 1-3	Expression
Reverse	AGGGACAGAAAAGCTGAGATGA	bp 147-566	NM_003226.3	control
MUC1				
Forward	TCCCAGCACCGACTACTACC	403bp	Exon 3-7	Expression
Reverse	CAGCTGCCCGTAGTTCTTTC	bp 300-702	NM_002456.5	variants 1-20
MUC2				
Forward	CTCCCAGACAGGAGAACGAG	227bp	Exon 24-26	Expression
Reverse	GAAGGTGACATAGTGCGGGT	bp 13277-13503	NM_002457.2	control
MUC4				
Forward	AAAACAGCCCACTGATGTCC	352bp Exon 8-10	NM_018406.5	Expression
Reverse	CAGCCTTCACGAAACTCTCC	bp 13623-13974	NM_004532.4	NM_138297.3
MUC6				
Forward	AGGCCTATGTCACTGTTGGG	440bp	Exon 13-16	Expression
Reverse	GATGGTGCAGTTGTCCACAC	bp 1598-2037	NM_005961.2	control
MUC5AC				
Forward	CTTCTTCAACACCTTCAAGACC	447bp	Exon 1-4	Expression
Reverse	AAGGTCTTGTAGTGGAAGTCAC	bp 1911-2357	XM_003119481.1	control
MUC5B				
Forward	CTTTGAGTACAAGAGAGTGGC	642bp	NM_002458.2	Expression
Reverse	TTTCATGATCAGACAATGCAC	bp 15097-15738	Exon 45-49	control

Appendix A

Clinical characteristics of 64 from 74 patients of the initial screening cohort (note that for 10 patients no clinical data was obtainable). TNM stands for tumor, nodes, metastasis of the UICC cancer staging system, G stands for histological grade, Loc for the location of the tumor (sigma, colon, rectosigmoid, colosigmoid or rectum) and type for either mucinous or nonmucious. The gender and age of the patient is also noted.

Pat. Nr.		T	N	M	G	Loc	Type	Age	Pat. Nr.		T	N	M	G	Loc	Type	Age
1	m	3	0	0	2	S	NM	63	33	f	4	2	1	1	C	NM	74
2	m	2	1	1	2	S	MU	79	34	m	3	0	1	2	S	NM	60
3	m	3	2	1	3	RS	MU	73	35	m	3	0	0	2	R	NM	69
4	f	3	0	0	3	R	NM	65	36	m	3	0	0	2	R	NM	66
5	f	3	0	0	2	C	MU	63	37	m	3	2	0	2	R	NM	76
6	m	4	2	0	2	S	NM	82	38	m	3	1	0	2	R	NM	60
7	m	3	0	0	2	R	MU	56	39	f	3	0	0	2	R	NM	61
8	m	4	0	0	2	C	NM	55	40	m	2	0	0	2	R	NM	76
9	f	4	0	0	2	R	NM	72	41	f	2	0	0	2	R	NM	58
10	m	2	0	0	2	C	NM	71	42	f	2	0	0	2	R	NM	79
11	f	3	1	0	3	RS	NM	67	43	f	2	0	0	2	C	NM	69
12	m	3	2	0	2	CS	NM	72	44	m	2	0	0	2	C	NM	68
13	m	4	0	0	1	C	MU	70	45	f	3	1	0	2	C	NM	77
14	m	3	2	0	3	R	MU	37	46	m	2	0	0	2	C	NM	68
15	m	2	1	0	3	C	MU	39	47	f	3	0	0	2	C	NM	83
16	f	3	1	0	2	CS	NM	57	48	f	4	2	0	2	C	NM	60
17	f	3	0	0	1	C	NM	85	49	m	2	0	0	1	S	NM	55
18	m	3	0	0	2	CS	NM	72	50	m	3	2	0	2	S	NM	59
19	m	3	0	0	1	C	NM	65	51	m	2	0	0	2	R	MU	67
20	f	3	0	0	3	C	NM	74	52	f	3	2	0	3	C	MU	86
21	f	3	0	0	2	C	NM	63	53	f	3	2	0	2	C	MU	66
22	m	4	2	0	3	S	MU	77	54	m	3	2	0	2	S	MU	93
23	f	2	0	0	2	S	NM	66	55	f	3	1	0	3	S	MU	83
24	m	3	1	0	2	C	NM	70	56	f	4	0	0	2	R	MU	77
25	m	3	1	0	3	RS	MU	72	57	m	4	2	0	3	S	MU	57
26	f	2	1	0	2	C	NM	60	58	f	3	0	0	3	R	NM	58
27	m	3	1	0	2	R	NM	66	59	m	3	1	0	2	R	NM	53
28	f	4	2	0	3	C	NM	60	60	m	4	1	0	2	C	NM	77
29	m	3	0	0	2	CS	NM	64	61	f	2	1	0	2	CS	NM	67
30	m	2	0	0	2	RS	NM	56	62	m	1	0	0	3	S	NM	67
31	f	2	2	0	3	R	NM	77	63	f	3	1	1	2	S	NM	60
32	f	3	2	0	3	R	NM	79	64	m	3	1	0	2	S	NM	59

Appendix A

Methylation levels of non-neoplastic mucosa and tumor samples (initial cohort of 74 patients).

TFAP2E

TFAP2E Pat. Nr.	PMR values NT	PMR values TU	Cutoff 30% Pos./Neg. NT/TU	Pat. Nr.	PMR values NT	PMR values TU	Cutoff 30% Pos./Neg. NT/TU
E1	25.13	39.54	P	28	11.74	41.58	P
E2	8.23	205.55	P	29	8.09	0.95	N
E3	7.71	35.99	P	30	32.92	63.43	N
E4	7.24	62.40	P	31	25.83	34.58	P
E5	243.32	197.98	N	32	47.84	13.48	N
E6	42.63	173.66	N	33	20.75	73.06	P
E7	21.36	10.01	N	34	46.51	20.66	N
E8	244.76	151.82	N	35	77.12	67.51	N
E9	56.40	168.00	N	36	71.66	63.83	N
E10	14.74	16.20	N	37	42.46	47.79	N
1	8.20	31.07	P	38	1.74	98.69	P
2	2.66	30.67	P	39	9.80	98.90	P
3	8.64	31.27	P	40	3.48	89.15	P
4	3.35	25.52	N	41	23.38	44.11	P
5	3.23	20.23	N	42	44.64	100.20	N
6	2.77	29.04	P	43	23.01	66.48	P
7	1.94	5.41	N	44	14.69	130.96	P
8	22.65	13.09	N	45	7.63	137.82	P
9	7.42	13.48	N	46	24.48	51.51	P
10	6.79	5.44	N	47	19.53	3.01	N
11	10.03	41.64	P	48	16.35	64.55	P
12	49.86	138.33	N	49	5.18	75.23	P
13	27.40	21.29	N	50	105.82	122.84	N
14	12.91	14.93	N	51	12.44	155.56	P
15	3.14	10.38	N	52	12.98	57.94	P
16	13.05	13.37	N	53	14.06	41.92	P
17	13.44	44.01	P	54	4.27	44.67	P
18	27.86	51.87	P	55	13.13	131.8	P
19	2.09	40.25	P	56	61.24	2.02	N
20	64.23	48.58	N	57	1.95	60.61	P
21	49.83	197.12	N	58	24.55	48.90	P
22	37.93	36.59	N	59	9.97	10.73	N
23	60.16	85.71	N	60	25.17	51.46	P
24	23.15	84.93	P	61	52.60	108.02	N
25	8.34	57.50	P	62	19.12	78.14	P
26	18.97	152.47	P	63	59.66	118.31	N
27	10.00	90.08	P	64	11.09	0.88	N

Appendix A

TUSC3

TUSC3 Pat. Nr.	PMR values NT	PMR values TU	Cutoff 30% Pos./Neg. NT/TU	Pat. Nr.	PMR values NT	PMR values TU	Cutoff 30% Pos./Neg. NT/TU
E1	19.12	59.57	P	28	66.76	63.99	N
E2	16.68	33.32	N	29	2.17	12.44	N
E3	1.44	1.82	N	30	44.38	59.39	P
E4	5.19	10.23	N	31	15.03	88.77	P
E5	0.77	57.10	P	32	27.73	16.55	N
E6	12.79	111.60	P	33	55.53	64.94	N
E7	0.48	2.27	N	34	68.31	58.44	N
E8	100.98	62.14	N	35	19.78	71.95	P
E9	96.62	285.52	N	36	28.51	15.87	N
E10	62.39	53.99	N	37	46.07	442.82	P
1	0.17	9.06	N	38	12.15	32.33	N
2	3.49	69.91	P	39	44.05	322.73	P
3	1.63	65.70	P	40	6.45	529.48	P
4	0.26	13.93	N	41	33.63	60.60	P
5	35.72	222.96	P	42	90.36	390.30	N
6	15.20	87.36	P	43	284.82	166.32	N
7	17.43	13.50	N	44	73.71	447.67	N
8	104.30	74.72	N	45	7.23	172.75	P
9	31.73	180.92	P	46	84.34	133.85	N
10	19.43	72.27	P	47	40.73	3.01	N
11	10.36	19.00	N	48	10.24	68.64	P
12	38.99	23.34	N	49	0.44	175.97	P
13	20.49	44.57	N	50	135.81	128.13	N
14	5.31	2.89	N	51	109.66	153.03	N
15	4.74	3.02	N	52	89.473	51.1987	N
16	3.77	7.88	N	53	51.899	170.248	P
17	8.55	2.01	N	54	47.002	114.168	P
18	13.08	31.89	N	55	68.916	190.85	N
19	2.47	18.40	N	56	124.5	160.326	N
20	11.45	18.45	N	57	57.44	92.2714	N
21	62.65	96.10	N	58	46.45	84.4795	P
22	64.43	49.95	N	59	42.515	155.396	P
23	22.56	15.04	N	60	235.77	108.178	N
24	46.77	96.51	P	61	45.645	211.741	P
25	8.64	6.21	N	62	60.356	288.038	N
26	169.61	288.78	N	63	39.941	289.614	P
27	81.56	71.53	N	64	57.666	25.0538	N

Appendix A

RXFP3

RXFP3 Pat. Nr.	PMR values NT	PMR values TU	Cutoff 30% Pos./Neg. NT/TU	Pat. Nr.	PMR values NT	PMR values TU	Cutoff 30% Pos./Neg. NT/TU
E1	0.17	226	P	28	4.06	48.87	Y
E2	91.72	82.44	N	29	2.65	1.59	N
E3	34.74	3.34	N	30	8.50	23.08	Y
E4	2.59	3.10	N	31	3.63	37.87	Y
E5	59.22	120.49	N	32	9.12	27.21	Y
E6	1.48	10.09	N	33	9.28	77.13	Y
E7	0.38	3.04	N	34	7.09	23.55	Y
E8	36.82	20.22	N	35	5.66	36.97	Y
E9	2.44	74.86	P	36	12.11	5.24	N
E10	5.86	5.17	N	37	6.17	126.86	Y
1	1.72	0.40	N	38	21.24	28.09	Y
2	1.41	0.63	N	39	2.50	70.52	Y
3	0.95	25.00	P	40	2.53	52.34	Y
4	1.52	42.31	P	41	5.92	33.00	Y
5	1.05	74.00	P	42	6.25	115.80	Y
6	0.00	0.71	N	43	3.87	71.07	Y
7	0.00	0.87	N	44	0.62	108.15	Y
8	4.00	46.00	P	45	0.44	81.60	Y
9	3.00	0.26	N	46	16.73	25.13	Y
10	0.45	3.67	N	47	18.32	3.46	N
11	1.00	111.03	P	48	9.51	54.54	Y
12	21.00	81.96	P	49	0.43	57.03	Y
13	0.01	24.00	P	50	97.20	128.72	N
14	1.16	3.24	N	51	9.19	31.178	Y
15	2.86	0.00	N	52	9.13	135.66	Y
16	2.39	25.95	P	53	5.60	102.29	Y
17	5.05	5.29	N	54	6.22	23.62	Y
18	14.23	43.07	P	55	10.00	10.13	N
19	7.09	150.56	P	56	21.67	7.92	N
20	15.98	18.25	N	57	8.50	49.04	Y
21	5.02	128.09	P	58	20.35	53.591	Y
22	3.07	87.85	P	59	7.107	6.4885	N
23	1.59	0.28	N	60	5.59	133.12	Y
24	4.31	24.18	P	61	4.961	149.89	Y
25	7.62	4.75	N	62	16.62	82.375	Y
26	21.97	87.46	P	63	5.97	5.00	N
27	5.07	8.96	N	64	7.92	3.77	N

Appendix A

List of Top 50 Up- and Downregulated genes found by expression microarray (Affymetrix Gene ST 1.0)
TFAP2E (SW480 cells) overexpressing clones vs. empty pTarget Vector control – downregulated in clones

FC	TOP 50 DOWN	Gene Family	(Pot.) Function
5.32	DKK4	dickkopf homolog 4 (Xenopus laevis)	WNT Signaling, up in CRC, promotes invasion, angiogenesis, via TCF/beta catenin
3.20	PARM1	prostate androgen-regulated mucin-like protein 1	Human PARM-1 is a novel mucin-like, androgen-regulated gene exhibiting proliferative effects in prostate cancer cells
2.52	PSG5	pregnancy specific beta-1-glycoprotein 5	up-regulated in cells circulating within blood from women with preeclampsia, and is positively correlated with clinical severity
2.48	ALDH1A3	Aldehyde dehydrogenase	Bile acid biosynthesis
2.40	PSG1	pregnancy specific beta-1-glycoprotein	support fetal survival and development
2.34	GABRA3	amma-aminobutyric acid receptor	ligand-gated chloride channel, neurotransmitter
2.24	CSGALNACT1	chondroitin sulfate N-acetylgalactosaminyltransferase 1	plays a role in the initiation and elongation in the synthesis of chondroitin sulfate
2.23	PSG7	pregnancy specific beta-1-glycoprotein 7	regulated by FOXF2 The reference genome contains a nonsense mutation that disrupts the coding sequence, suggesting that this gene may be evolving into a pseudogene
2.15	BNC2	basonuclin	regulatory protein of DNA transcription
2.14	NEDD9	neural precursor cell expressed, developmentally down-regulated	cell cycle, adhesion, growth, cytoskeleton, integrin + cytokine signaling, inflammation, TGFbeta, NFkappaB
2.09	FGF9	fibroblast growth factor	Sonic hedgehog (Shh) signaling, beta catenin, cell cell signals, Wnt pathway, VEGF pathway, proliferation, cell growth, signal transduction, MAPKKK, angiogenesis
2.07	GPR155	G protein-coupled receptor	UNKNOWN
2.06	RNPC3	RNA-binding region	component of the U12-type spliceosome
2.01	TNFRSF11B	tumor necrosis factor receptor	negative regulator of bone resorption, osteoclast development, lymph-node organogenesis and vascular calcification, Cytokines
2.01	CST1	cystatin SN	cysteine protease inhibitor
1.98	IGFL4	IGF-like family	growth and development
1.96	THBS1	thrombospondin	cell-to-cell and cell-to-matrix interactions, angiogenesis, tumorigenesis
1.94	OR6B2	olfactory receptor	GPR, signal transduction
1.93	CYP4A11	cytochrome	PPAR signaling pathway, drug metabolism and synthesis of cholesterol, steroids and other lipids, hydroxylates medium-chain fatty acids such as laurate and myristate
1.93	LRP2	low density lipoprotein	Heymann nephritis antigenic complex, Hedgehog signaling pathway, cell proliferation
1.93	CDH11	Cadherin	cell-cell adhesion, osteoblastic, TNFalpha induced, metastasis
1.91	TNC	tenascin C	associated with metastasis in clear cell renal cell carcinoma, adhesion, cell cell signaling, ecm interaction, survival, TGFbeta induced, EGF, proliferation, progression
1.88	ABCB1	ATP-binding cassette	multidrug resistance
1.88	KITLG	KIT ligand	cell migration
1.87	CA8	carbonic anhydrase	promotes colon cancer cell growth
1.86	SPRR2D	small proline-rich	keratinization, keratinocyte differentiation
1.86	GPR52	G protein-coupled receptor	signal transduction from the external environment to the inside of the cell
1.86	PSG11	pregnancy specific beta-1-glycoprotein	The human pregnancy-specific glycoproteins (PSGs) are a group of molecules that are mainly produced by the placental syncytiotrophoblasts during pregnancy. PSGs comprise a subgroup of the carcinoembryonic antigen (CEA) family, which belongs to the immunoglobulin superfamily.
1.85	SNAI2	snail homolog	C2H2-type zinc finger transcription factor, transcriptional repressor that binds to E-box motifs and is also likely to repress E-cadherin transcription in breast carcinoma.

			This protein is involved in epithelial-mesenchymal transitions and has antiapoptotic activity, epithelial to mesenchymal transition.
1.84	SAMD9	sterile alpha motif domain containing	cell proliferation, TNFalpha signaling
1.84	UGT2B10	UDP glucuronosyltransferase	nicotine metabolism
1.82	MUCL1	mucin-like	breast cancer micrometastasis
1.82	GJA5	gap junction membrane channel protein	Cell Communication
1.80	ANKRD36B	ankyrin repeat domain 36B	melanoma-associated and CLL antigen
1.80	VAT1L /KIAA1576	vesicle amine transport protein	oxidation reduction
1.80	ANXA1	annexin A1	anti-inflammatory activity
1.79	TAS2R44	taste receptor	GPR, signal transduction
1.78	ENTPD8	ectonucleoside triphosphate diphosphohydrolase	main hepatic NTPDase activity
1.78	PNPLA7	patatin-like phospholipase domain	adipocyte differentiation
1.77	SERPINI1	serpin peptidase inhibitor	regulation of axonal growth and the development of synaptic plasticity
1.77	DIRAS2	inhibits tissue-type plasminogen activator	DIRAS2 belongs to a distinct branch of the functionally diverse Ras (see HRAS; MIM 190020) superfamily of monomeric GTPases.
1.73	UTRN	utrophin	actin-binding, muscle development,The protein encoded by this gene is located at the neuromuscular synapse and myotendinous junctions, where it participates in post-synaptic membrane maintenance and acetylcholine receptor clustering.
1.72	DCLK1	doublecortin-like kinase	kinase, microtubule polymerizing activity, differentiation, axon extension
1.71	PPME1	protein phosphatase methylesterase	catalyzes the demethylation of the protein phosphatase-2A catalytic subunit, Alkaloid biosynthesis
1.71	PLA2G10	phospholipase	anti-inflammatory response
1.70	HNT/RREB1	ras responsive element binding protein	reduce cell-cell adhesion
1.70	LRLE1	liver-related low express protein	UNKNOWN
1.69	PPM2C	protein phosphatase	dephosphorylation and reactivation of the pyruvate dehydrogenase complex
1.69	GPR21	G protein-coupled receptor	UNKNOWN
1.68	SAA2	serum amyloid A2	Increased expression of SAA2 by adipocytes in obesity may play a critical role in local and systemic inflammation and free fatty acid production and could be a direct link between obesity and its comorbidities.

TFAP2E overexpressing clones versus empty pTarget Vector control – upregulated in TFAP2E clones

FC	TOP 50 UP	Gene Family	(Pot.) Function
23.97	MAGEC2	melanoma antigen family	cancer, promoter demethylated by Kit
8.65	ASB4	ankyrin repeat and SOCS box-containing	cytokine signalling, energy homeostasis
3.12	MUC13	MUCINS	down in CRC, inflammation, differentiation, reflux
2.82	SYTL3	synaptotagmin-like	cell signaling ?, intracellular protein transport
2.44	SLC1A2	solute carrier family	glutamate transporter, extracellular signaling
2.43	VIL1	calcium-regulated actin-binding	Differentiation, cytoskeleton, cell migration
2.39	GMFG	glia maturation factors	stem cells, kinase inhibitor, glycosylation
2.36	CKMT2	creatine kinase	metabolism, energy transfer mitochaondria
1.98	CKMT1A	creatine kinase	metabolism, energy transfer mitochaondria
1.98	DDIT4	DNA-damage-inducible	DNA-damage repair, mTOR Signaling, PI3K
1.97	SERPINE1	serpin peptidase inhibitor	angiogenesis, TNFalpha, TGFbeta, invasion, H pylori, apoptosis
1.93	TIGD2	tigger transposable	transposon element
1.91	GRB10	growth factor receptor-bound	inhibits tyrosine kinase, suppression of growth, apoptosis, insulin signaling

Appendix A

1.86	PTPN13	tyrosine phosphatase	cell growth, differentiation, mitotic cycle, oncogenic transformation, Fas mediated programmed cell death, via GTPase Rho signaling
1.85	SLC7A7	solute carrier family	cationic amino acid transporter
1.84	EPS8L3	EGFR like	substrate for the epidermal growth factor receptor
1.84	CTH	cystathionase	trans-sulfuration pathway, Cysteine metabolism
1.83	PRAME	preferentially expressed antigen in melanoma	antigen that is predominantly expressed in human melanomas and that is recognized by cytolytic T lymphocytes. It is not expressed in normal tissues, except testis; expression pattern is similar to that of other CT antigens, such as MAGE, BAGE and GAGE
1.83	UPP1	uridine phosphorylase 1	converting 5'dFUrd/5FU into 5FU/Furd, prognostic marker in human breast carcinoma, regulated by p53, associated with lymph node metastasis in oral neoplasms
1.83	PIP5K1B	hosphatidylinositol-4-phosphate 5-kinase	Regulation of actin cytoskeleton, cell signaling
1.82	NT5E	nucleotidase	marker of lymphocyte differentiation
1.81	HNF4A	hepatocyte nuclear factor	p38 signaling pathway, bile acid synthesis, downregulation of proliferation, cell growth
1.79	CACNB4	calcium channel	MAPK signaling pathway, modulating G protein inhibition, increasing peak calcium current,
1.78	CDGAP	GTPase-activating protein	protein trafficking and cell growth, apoptose
1.77	DDIT3	DNA-damage-inducible	MAPK signaling pathway, p38 induced, negatively regulate cell growth and survival
1.76	GEM	GTP binding protein	overexpressed in skeletal muscle, regulatory protein in receptor-mediated signal transduction
1.75	FCGR2A	immunoglobulin Fc receptor	immune response, phagocytosis and clearing of immune complexes
1.74	ERVWE1	endogenous retroviral family	important in reproduction
1.73	E2F5	E2F transcription factor	control of cell cycle and action of tumor suppressor proteins, interacts with p130 and p107
1.72	ZNF578	zinc finger protein	UNKNOWN
1.72	MTERFD1	mitochondrial transcription termination factor	UNKNOWN
1.72	ATP1A3	ATPase, Na+/K+ transporting	electrochemical gradient
1.72	MT1A	metallothionein 1A	cell proliferation, diabetes mellitus 2, cell viability, apoptosis, ovarian cancer, heavy metal detoxification, chemoresistance
1.72	ZFAND1	zinc finger, AN1-type domain 1	UNKNOWN
1.72	THNSL1	threonine synthase-like	UNKNOWN
1.71	MIA	melanoma inhibitory activity	melanoma tumor progression, lymph node metastasis, immunosuppression, matrix adhesion
1.7	CLIC5	chloride intracellular channel	ion transport
1.7	TM4SF19	transmembrane 4 L	UNKNOWN
1.69	NEK3	NIMA (never in mitosis gene a)-related kinase	mitotic regulation, polymorphism associated with prostate cancers, breast cancer motility
1.69	RSPO3	R-spondin 3 homolog	signal transduction by receptor tyrosine kinases
1.69	HKDC1	hexokinase domain	Glycolysis
1.68	PDE3B	phosphodiesterase	Insulin signaling, cAMP, Apoptosis
1.68	HRASLS	HRAS-like suppressor	UNKNOWN
1.67	TATDN1	TatD DNase domain	hepatocarcinoma high expression;putative deoxyribonuclease
1.67	FEZ1	asciculation and elongation protein	cell adhesion, axon guidance
1.67	ALS2CR2/STRADB	STE20-related kinase adaptor	pseudokinase, component of a complex involved in the activation of serine/threonine kinase 11, enhances the anti-apoptotic activity of this protein via the JNK1 signal transduction pathway.
1.66	FAM175A	UNKNOWN	DNA damage response and repair, breast cancer susceptibility
1.66	TTC30B	tetratricopeptide repeat domain	UNKNOWN
1.65	BST2	bone marrow stromal cell antigen	role in pre-B-cell growth and in rheumatoid arthritis, Interferon-alpha enhances BST2 expression, nfKappab
1.65	ADFP	adipose differentiation-related protein	renal carcinoma differentiation, protective effects against apoptosis, fatty acids, PPAR

TUSC3 overexpressing clones versus empty pTarget Vector controls – downregulated in clones (SW480 cells)

FC	TOP 50 DOWN	Gene Family	(Pot.) Function
5.17	PSG7	pregnancy specific beta-1-glycoprotein 7	downregulated by FOXF2, stroma marker in prostate

Appendix A

5.03	PSG5	pregnancy specific beta-1-glycoprotein 5	rxr-alpha regulated RARE element in promoter and KLFs
4.43	KIAA1199	KIAA1199	overexpressed in GC and CRC, WNT
4.11	CYP24A1	cytochrome P450, family 24, subfamily A, polypeptide 1	considered main enzyme VitD half life
4.09	PSG1	pregnancy specific beta-1-glycoprotein 1	immunomodulation, liver, CEA subgroup
4.08	ITGB8	integrin, beta 8	metastasis prediction, TGFß regulated
4.07	SEMA5A	sema domain, seven thrombospondin repeats (type 1 and type 1-like), transmembrane domain (TM) and short cytoplasmic domain, (semaphorin) 5A	invasion migration metastasis angiogenesis
3.87	TNC	tenascin C	prognostic marker, invasion, angiogenesis
3.60	SCNN1A	sodium channel, nonvoltage-gated 1 alpha	cystic fibrosis, retinoic acid
3.56	ANKRD36B	ankyrin repeat domain 36B	no info+
3.51	FAT1	FAT tumor suppressor homolog 1 (Drosophila)	migration
3.07	MUC21	mucin 21, cell surface associated	up in cervical adenocarcinoma
3.03	CDH1	cadherin 1, type 1, E-cadherin (epithelial)	proliferation, invasion, and/or metastasis
3.03	TNFSF10	tumor necrosis factor (ligand) superfamily, member 10	induces apoptosis in tumor cells
3.02	CLIP4	CAP-GLY domain containing linker protein family, member 4	no info+
3.02	MAMDC2	MAM domain containing 2	Kabuki syndrome.
2.97	CD24	CD24 molecule	prognostic marker, metastasis, migration, invasion, CSCs
2.96	UTRN	utrophin	muscle dystrophy
2.91	ZNF594	zinc finger protein 594	no info+
2.87	TGFB2	transforming growth factor, beta 2	overexpressed in crc, stromal cells, fibroblasts
2.79	GPRC5A	G protein-coupled receptor, family C, group 5, member A	RAR/RXR woundhealing, mucosa
2.79	hsa-mir-23b	**MICRO-RNA**	stemness factor CD34(+)CD38(-)
2.73	OVOS	ovostatin	no info+
2.73	FN1	fibronectin 1	marker for node-positive CRC, metastasis
2.69	TNFRSF11B	tumor necrosis factor receptor superfamily, member 11b	methylated in crc/up in crc apoptose resistance, migration, adhesion
2.68	FBXO32	F-box protein 32	muscle atrophy
2.64	ZNF487	zinc finger protein 487	no info+
2.57	RAB3B	RAB3B, member RAS oncogene family	regulation of exocytosis, regulation of pituitary hormone secretion
2.53	LCN2	lipocalin 2	up in aggressive rectal cancer, stress resistance, heliobacter, cell adhesion

Appendix A

FC		Gene Family	(Pot.) Function
2.53	SYNE2	spectrin repeat containing, nuclear envelope 2	actin cytoskeleton connection
2.50	RUNX1	runt-related transcription factor 1	regulates FOXP3
2.49	OVOS2	ovostatin 2	no info+
2.46	AMOT	angiomotin	angiogenesis, migration
2.46	TMC4	transmembrane channel-like 4	skin cancer
2.43	CYR61	cysteine-rich, angiogenic inducer, 61	tgfß, lymphnode metastasis
2.43	S100A3	S100 calcium binding protein A3	tnm stage in GC, metastasis
2.42	FAM111B	family with sequence similarity 111, member B	no info
2.41	SIGLEC6	sialic acid binding Ig-like lectin 6	muccosa in lymphomas
2.40	E2F7	E2F transcription factor 7	dna damage response, platinum resistance in ovarian cancer
2.38	PHLDB2	pleckstrin homology-like domain, family B, member 2	cytoskeleton
2.36	MFI2	antigen p97 (melanoma associated) identified by monoclonal antibodies 133.2 and 96.5	osteoblasts
2.35	hsa-mir-24-1	**MICRO-RNA**	apoptosis, inhibits proliferation by targeting MYC, E2F2
2.32	ARPP-19	cyclic AMP phosphoprotein, 19 kD	cAmp
2.32	WEE1	WEE1 homolog (S. pombe)	dna damage
2.30	FOSB	FBJ murine osteosarcoma viral oncogene homolog B	oxidative stress
2.30	GABARAPL1	GABA(A) receptor-associated protein like 1	triggered by FOXO, chaperone
2.29	PSG11	pregnancy specific beta-1-glycoprotein 11	immunomodulator, TGFß, IL10+IL6
2.28	hsa-mir-27b	**MICRO-RNA**	differentiation of endometrial stromal cells
2.25	HUNK	hormonally up-regulated Neu-associated kinase	breast cancer metastasis, prognostic and diagnostic marker
2.25	SLC16A13	solute carrier family 16, member 13 (monocarboxylic acid transporter 13)	downregulated by PPARalpha

TUSC3 overexpressing clones versus empty pTarget Vector controls – upregulated in clones (SW480 cells)

FC	TOP 50 UP	Gene Family	(Pot.) Function
10.88	SPARC	secreted protein, acidic, cysteine-rich (osteonectin)	scaffold, vitamin d
5.74	HSPA6	heat shock 70kDa protein 6 (HSP70B')	heat shock, stress response, COX2, myc/myb regulated
3.82	ASB4	ankyrin repeat and SOCS box-containing 4	damage response?
3.28	CD300A	CD300a molecule	inflammatory response
3.26	CCR7	chemokine (C-C motif) receptor 7	lymph nodes, metastasis prediction marker
2.92	UIMC1	ubiquitin interaction motif containing 1	damage response, overexpressed in tumor
2.92	COL6A3	collagen, type VI, alpha 3	inflammation response, platin resistance, Cox1
2.91	GEM	GTP binding protein overexpressed in skeletal muscle	immune response

Appendix A

2.77	GPR35	G protein-coupled receptor 35	diabetes, overexpressed in GC
2.72	**NUPR1**	nuclear protein 1	stress response
2.63	VIL1	villin 1	lymph node marker
2.60	MUC13	mucin 13, cell surface associated	overexpressed in crc and gc
2.58	LAPTM5	lysosomal multispanning membrane protein 5	biomarker lung cancer
2.55	SLAIN1	SLAIN motif family, member 1	stem cell marker
2.46	MSL3L2	male-specific lethal 3-like 2 (Drosophila)	binds collagens
2.41	PRRX1	paired related homeobox 1	hypoxia
2.35	RASGRF1	Ras protein-specific guanine nucleotide-releasing factor 1	promote (GDP)/(GTP) exchange, inflammation, MMPs, imprinted
2.30	HNF4A	hepatocyte nuclear factor 4, alpha	ulceritis, diabetes, intestinal gc, metastasis
2.25	POLR3G	polymerase (RNA) III (DNA directed) polypeptide G (32kD)	no info
2.22	TOMM22	translocase of outer mitochondrial membrane 22 homolog (yeast)	interaction with HSP70 and CYPs
2.22	ZEB1	zinc finger E-box binding homeobox 1	EMT, drug resistance, migration, downregulation of miRNAs, marker for metastasis and poor survival, regulates VDR
2.21	USH1C	Usher syndrome 1C (autosomal recessive, severe)	induces G2/M phase cell cycle arrest
2.19	GPD1L	glycerol-3-phosphate dehydrogenase 1-like	hypoxia, sodium channels
2.14	OR7D4	olfactory receptor, family 7, subfamily D, member 4	no info
2.08	RIOK1	RIO kinase 1 (yeast)	Myc-associated protein trhough MAPJD
2.07	VGF	VGF nerve growth factor inducible	neuroendocrine neoplasia
2.06	RAB39B	RAB39B, member RAS oncogene family	overexpressed in germ cell neoplasia
2.05	DOC2B	double C2-like domains, beta	SNARE regulator of glucose-stimulated delayed insulin secretion, vesicle transport
2.04	NOPE	neighbor of Punc E11	cell surface markers for murine fetal hepatic stem cells, **methylated** in lymphoblastic leukemia
2.02	TUBE1	tubulin, epsilon 1	required for centriole duplication and microtubule organization
2.01	hsa-mir-320a	**MICRO-RNA**	associated with tumorigenesis of retinoblastoma, Diagnostic and prognostic microRNAs in stage II colon cancer
2.00	hsa-mir-124-2	**MICRO-RNA**	epigenetically silenced tumor-suppressive microRNAs in hepatocellular carcinoma
1.98	MEFV	Mediterranean fever	important modulator of innate immunity
1.97	CA9	carbonic anhydrase IX	hypoxia.
1.97	DLEU1		deleted in lymphocytic leukemia 1
1.96	NAT8		N-acetyltransferase 8
1.95	SLC7A7		solute carrier family 7 (cationic amino acid transporter, y+ system), member 7
1.93	EIF4B		eukaryotic translation initiation factor 4B
1.93	NSBP1		nucleosomal binding protein 1
1.93	SP140L		SP140 nuclear body protein-like
1.93	TRIB3		tribbles homolog 3 (Drosophila)
1.93	GDF7		growth differentiation factor 7
1.92	SNORA70C		small nucleolar RNA, H/ACA box 70C
1.91	GRB10		growth factor receptor-bound protein 10
1.91	DENR		density-regulated protein
1.91	LOC390414		hypothetical LOC390414
1.91	PPID		peptidylprolyl isomerase D (cyclophilin D)
1.90	ZFAND1		zinc finger, AN1-type domain 1

RXFP3 overexpressing clones versus empty pTarget Vector controls – upregulated in clones (SW480 cells)

Foldchange	Gene	Function
5.48	ROCK1	CpGI, cell adhesion, actin cytoskeleton, PTEN, TGFB, Apoptosis, insulin resistance, invasion, angiogenesis, maybe **hypoxia** related
4.32	**SCNN1A**	CpGI, DNMT3b regulated, cyxtic fibrosis, homoestasis, stress, actin, repressed by Retinoic acid
4.21	LRRFIP1	CpGI, represses TNF, glioblastoma

Appendix A

4.09	**DDIT4**	hypoxia,Up by p53 and BRCA1,Gas5FUresDN, stress response, chemotherapy, DNA damage, apoptosis
4.01	PDGFA	linked to **HIF1A**,proliferation, oxidative damage
3.70	DKK4	
3.46	**CA9**	**hypoxia**, hif1 target, ,resistance to chemotherapy, poor prognosis, metastasis
3.43	KCTD11	hedgehog suppressor
3.35	7A5 (MACC1)	metastasis-associated in colon cancer, prognostic indicator of metastasis formation
3.34	ANKRD37	**hypoxia**
3.27	NRN1	**hypoxia**
3.09	GDF15	Up-regulated at 24 hours following treatment of human lymphocytes (TK6) with a high dose of methyl methanesulfonate (MMS)
3.00	EGLN3	CpGI, **hypoxia** protective, HIF1A loop
3.00	CARD11	NFKB, IKK
2.99	MOBKL2A	no info
2.97	SC4MOL	response to interleukin-3 - downregulated in differentiation pathways,Androgen regulated genes in the murine epididymis, downregulated by mutations in EGFR2,Up-regulated in mouse liver tissue from mice with reduced liver expression of NADPH-cytochrome P450 reductase
2.95	BNIP3L	**hypoxia**, proapoptotic
2.90	SNORD3A	
2.83	INSIG1	**hypoxia**
2.78	ATP11A	predictive marker for metastasis
2.77	FGD5	actin
2.73	IRF2	interefon, myc
2.71	SCD	downregulated in differentiation pathways, hypoxia, downregulated by mutations in EGFR2insulin resistance, inflammation, prostate cancer
2.70	TMCO3	mir-506 target
2.59	NDRG1	**hypoxia**, myc, proliferation, stress response, differentiation
2.58	CSRP2	hypoxia, Up by p53 and BRCA1 and TGFB, Lim related, methylated in mouse skin cancer
2.53	WDR74	up in lung metastasis of breast cancer
2.46	ACOT8	peroxisomes
2.45	CAPG	hypoxia, actin cytoskeleton-associated, breast cancer metastasis, prognostic factor for ovarian carcinoma, tsg, methylated
2.44	PTGES	**hypoxia, NO**, ICOSANOID_METABOLIC_PROCESS, inflammatory response, **nitric oxide, crc**
2.43	ERRFI1	**stress response**, downregulated by mutations in EGFR2
2.41	DPYSL2	crc marker
2.37	KCTD11	hedgehog signaling, tsg
2.35	DUX4	sensitive to oxidative stress - hypermethylated in cervical cancer
2.34	DHRS3	OXIDOREDUCTASE_ACTIVITY, methylated in melanoma, lymph node metastasis
2.32	SORL1	alzheimer
2.32	CDH1	hypoxia,actin cytoskeleton-associated,CELL_JUNCTION,INTERCELLULAR_JUNCTION
2.29	ZAP70	
2.26	CLDN7	CELL_JUNCTION,INTERCELLULAR_JUNCTION,CELL_CELL_ADHESION
2.25	IDH2	hypoxia, brca, breast cancer
2.25	CORO1A	actin cytoskeleton-associated
2.25	IGF1R	**oxidative stress**
2.21	OSBPL5	downregulated by mutations in EGFR2
2.18	RBM35A	tumor suppressor in CRC, MIR-519E target
2.17	ALDH3A2	downregulated by telomerase,OXIDOREDUCTASE_ACTIVITY, peroxisome
2.15	ADM	**hypoxia, hif1 target**
2.15	GSN	actin cytoskeleton-associated
2.13	IGFBP3	**hypoxia, hif1 target**
2.12	CHRNB1	ACETYLCHOLINE_BINDING
2.12	STARD4	**hypoxia**
2.12	FBXO2	ubiqutin protein ligase
2.10	CDH3	cell adhesion, cervical cancer, demethylated in gastric cancer, BRCA target

Appendix A

2.10	MFAP2	antigen of elastin-associated microfibrils
2.07	C8orf73	validated protein coding
2.04	CLDN1	CELL_JUNCTION,INTERCELLULAR_JUNCTION,CELL_CELL_ADHESION
2.03	NRGN	schizophrenia
2.02	FUCA1	HDAC regulated, fucosidosis
2.02	PMS2L1	pseudogene
2.02	DHCR7	Up-regulated in mouse liver tissue from mice with reduced liver expression of NADPH-cytochrome P450 reductase
2.02	CYFIP2	CELL_CELL_ADHESION ; ALL
2.01	TMEM163	
2.01	SECTM1	immune response
2.01	C17orf76	validated, mir target
2.00	HOXC5	breast cancer

RXFP3 overexpressing clones versus empty pTarget Vector controls – downregulated in clones (SW480 cells)

Foldchange	Gene	Function
13.27	P2RY5	other name LARP6 colitis, bladder cancer, ovarian cancer, embedded in RB1
6.25	ZFAND2A	nfkb target, copper regulated, CpGI, stress response
4.85	MAGEC2	no CpGI but can be methylated, KIT regulated, lung and melanoma, eso
3.57	DKK1	Upregulated by H2O2, Menadione and t-BH in breast cancer cells
3.40	ARRDC4	regulates txnip
3.27	SNAI2	HOXA5_TARGET,Down-regulated in fibroblasts following infection with human cytomegalovirus, Vitamin D
3.16	PPP1R15A	**hypoxia in astrocytes**
3.04	DDIT3	Upregulated by H2O2, Menadione and t-BH in breast cancer cells, methylated in ML, oxidative stress
2.96	EFCAB4B	no info
2.78	TXNIP	Genes downregulated in response to glutamine starvation, **hypoxia**, inflammation
2.63	BIRC3	HOXA5_TARGET,Genes up-regulated by TNFA in colon,derm,iliac,aortic,lung endothelial cells
2.62	SERPINE2	lymph node metastasis, diabetes
2.62	HSPA1A	
2.62	NRIP1	Up-regulated at any timepoint up to 24 hours following infection of HEK293 cells with reovirus strain T3Abney,Down-regulated consistently at 6-24 hours following treatment of WS1 human skin fibroblasts
2.58	ZNF587	mir-370 target
2.57	POM121C	
2.56	HBEGF	Regulated by UV-B light in normal human epidermal keratinocytes, Up-regulated in fibroblasts following infection with human cytomegalovirus
2.51	HSPA1B	
2.46	ND2	mitochondrial, glucose tolerance
2.42	BHLHB9	methylated in crc
2.39	KRT81	promotes breast cancer lung metastasis
2.34	CENPBD1	hypomethylated promoters upregulated by the combination of TSA and DAC in ovarian carcinoma (CP70) cells
2.34	RRS1	Genes downregulated in human pulmonary endothelial cells under hypoxic conditions or after exposure to AdCA5,Down-regulated in fibroblasts following infection with human cytomegalovirus,Genes up-regulated by MYC in P493-6 (B-cell),Genes downregulated in response to glutamine starvation , resistance gene
2.31	NFKBIA	Genes up-regulated by TNFA in colon,derm,iliac,aortic,lung endothelial cells,HYPOXIA_NORMAL_UP
2.31	FAM131C	
2.30	HSPA6	MYC_ONCOGENIC_SIGNATURE
2.26	PSG6	interleukin inducible
2.25	GPN2	
2.23	ZNF57	suppresses NFAT and p21 pathway
2.21	MRPS25	target of micrornas 33, 512-3P, 320
2.18	GJC1	
2.17	C17orf91	validated, mir target
2.17	C18orf45	mir 18a target

Appendix A

2.16	ZDHHC14	leukemia
2.16	AASS	down by sulindac
2.15	CAMKK2	regulation of energy balance
2.15	MOSPD1	target of mirs
2.13	CYP27B1	methylated in ovarian carcinoma, vitamin d related, colon and prostate cancer risk
2.12	PLK2	upregulated with TNFa treatment, only with functional NFkB
2.10	BLMH	Up-regulated in fibroblasts following infection with human cytomegalovirus
2.10	GADD45B	HOXA5_TARGET
2.10	FAM126B	Down-regulated in glomeruli isolated from Pod1 knockout mice, versus wild-type controls
2.09	JUN	HOXA5_TARGET,Upregulated by H2O2, Menadione and t-BH in breast cancer cells
2.08	NSF	Genes on chromosome 17 with copy-number-driven expression in pancreatic adenocarcinoma.
2.08	AP1GBP1	mir target
2.07	LAD1	regulated by tsa, sulindac, butyrate, anchoring element of the basement membrane
2.07	TBX3	HOXA5_TARGET, up in breast cancer
2.06	SRFBP1	interacts with HSPs
2.05	DIO3	regulation of thyroid hormone inactivation
2.05	GPATCH4	multiple myeloma, hcc
2.05	RND3	Up-regulated at any timepoint up to 24 hours following infection of HEK293 cells with reovirus strain T3Abney,Regulated by UV-B light in normal human epidermal keratinocytes
2.03	CPLX2	mir target, neurotransmitter
2.03	SGK269	protein kinase, reacts to growth hormone
2.03	ORC6L	Genes downregulated in human pulmonary endothelial cells under **hypoxic** conditions or after exposure to AdCA5, an adenovirus carrying constitutively active hypoxia-inducible factor 1 (HIF-1alpha).
2.03	CRYAB	Up by H2O2, up by MKK6, negative regulation of intracellular transport, p53 target, HSP27 interaction
2.03	FEM1A	up by sulindac, polycystic ovary syndrome
2.01	SRRD	Sox2 enhancer, epigenetically regulated
2.01	UIMC1	
2.01	C17orf58	validated, mir492, breast cancer
2.01	ANXA1	Upregulated by dsRNA (polyI:C) in IFN-null GRE cells,Up-regulated at any timepoint up to 24 hours following infection of HEK293 cells with reovirus strain T3Abney
2.00	WDR73	
2.00	C17orf70	dna damage response, mir target 128 und 29

Appendix B

TUSC3 Migration and Adhesion Data

5000 Cells	Matrigel	Collagen	Control	% Invasion	%Adhesion
K30	102	330	300	34	110
K31	108	309	300	36	103
K33	114	321	300	38	107
PTM	130	240	250	52	96
P4	130	270	250	52	108
Mean K	108	320	300	36	106.67
Mean P	130	255	250	52	101.33
		Invasion Index	1.44	Adhesion Index	0.95
10 000 Cells	**Matrigel**	**Collagen**	**Control**	**% Invasion**	**%Adhesion**
K33	101	400	340	29.7	117.65
K34	117	370	364	32.1	101.65
K35	118	323	360	32.8	89.72
PT2	164	440	344	47.7	127.91
PT3	135	480	340	39.7	141.18
Mean K	112.00	364.33	354.67	31.54	103.01
Mean P	149.50	460.00	342.00	43.69	133.03
		Invasion Index	1.39	Adhesion Index	1.29
Mean K	110.00	342.17	327.33	33.77	104.84
Mean P	139.75	357.50	296.00	47.85	117.18
		Invasion Index	1.42	Adhesion Index	1.12

TFAP2E

Expression Data for 28 CRC patient samples (TU) for **TFAP2E** and **DKK4**

Patient no.		TFAP2E Methylation	TFAP2E Expression	DKK4 expression	Patient no.		TFAP2E Methylation	TFAP2E Expression	DKK4 expression
		tumor	tumor	tumor			tumor	tumor	tumor
49	TFAP Methylation		no	yes	13	TFAP not methylated		no	yes
26			no	yes	34			no	no
57			no	yes	E10			no	no
18			no	yes	16			no	no
45			no	yes	56			no	no
24			no	yes	22			no	no
51			no	yes	20			yes	no
31			no	yes	42			yes	yes
55			no	no	12			no	yes
54			no	no	30			yes	yes
28			no	no	8			no	no
27			yes	no					
25			yes	yes					
38			yes	yes					
39			yes	yes					
19			yes	yes					
40			no	yes					

Appendix B

Ranplex Mutiations

TP53	K-ras	APC		BRAF
175.2RH	12.1GR	876.1	1367.1	600.2VE*
245.1GS	12.1GC	1306.1	1378.1	
245.2GV	12.1GS	1309.1	1379.1	
248.1RW	12.2GA	1309.5del	1450.1	
248.2RQ	12.2GD	1312.1	1465.2del	
273.1RC	12.2GV	1338.1	1554.1ins	
273.2RH	13.2GD			
282.1RW		* Previously BRAF 599.2 VE		

Explanation

TP53	Codon	normal AA	mutated AA
175.2RH	175	Arginine	Histidine
245.1GS	245	Glycine	Serine
245.2GV		Glycine	Valine
248.1RW	248	Arginine	Tryptophan
248.2RQ		Arginine	Glutamine
273.1RC	273	Arginine	Cysteine
273.2RH		Arginine	Histidine
282.1RW	282	Arginine	Tryptophan
KRAS	Codon	normal AA	mutated AA
12.1GR	12	Glycine	Arginine
12.1GC		Glycine	Cysteine
12.1GS		Glycine	Serine
12.2GA		Glycine	Alanine
12.2GD		Glycine	Aspartic Acid
12.2GV		Glycine	Valine
13.2GD	13	Glycine	Aspartic Acid
BRAF			
600.2VE	600	Valine	Glutamic Acid
APC	Codon	normal AA	mutated AA
876.1			nonsense
1306.1			nonsense
1309.5del			deletion of 5 bases
1309.1			nonsense
1312.1			nonsense
1338.1			nonsense
1367.1			nonsense
1378.1			nonsense
1379.1			nonsense
1450.1			nonsense
1465.2del			deletion of 2 bases
1554.1ins			insertion of 1 base

Appendix B

Data from Ranplex CRC array

Sample	Tissue	RanplexCRC Markers
1	normal	K_RAS12_1WT, BRAF600_2WT, TP53175_2WT, APC876_1WT, APC1450_1WT
P 16	tumour	K_RAS12_1WT, BRAF600_2WT, TP53175_2WT, APC876_1WT, APC1450_1WT
2	normal	**BRAF600_2**, K_RAS12_1WT, BRAF600_2WT, TP53175_2WT, APC876_1WT, APC1450_1WT
P 17	tumour	**BRAF600_2**, K_RAS12_1WT, BRAF600_2WT, TP53175_2WT, APC876_1WT, APC1450_1WT
3	normal	K_RAS12_1WT, BRAF600_2WT, TP53175_2WT, APC876_1WT, APC1450_1WT
P 18	tumour	K_RAS12_1WT, BRAF600_2WT, TP53175_2WT, APC876_1WT, APC1450_1WT
4	normal	K_RAS12_1WT, BRAF600_2WT, TP53175_2WT, APC876_1WT, APC1450_1WT
P 19	tumour	**K-RAS12_1SE**, K_RAS12_1WT, BRAF600_2WT, TP53175_2WT, APC876_1WT, APC1450_1WT
5	normal	K_RAS12_1WT, BRAF600_2WT, TP53175_2WT, APC876_1WT, APC1450_1WT
P 20	tumour	**K-RAS12_1SE**, K_RAS12_1WT, BRAF600_2WT, TP53175_2WT, APC876_1WT, APC1450_1WT
6	normal	K_RAS12_1WT, BRAF600_2WT, TP53175_2WT, APC876_1WT, APC1450_1WT
P 21	tumour	K_RAS12_1WT, BRAF600_2WT, TP53175_2WT, **APC1309_5del**, APC876_1WT, APC1450_1WT
7	normal	K_RAS12_1WT, BRAF600_2WT, TP53175_2WT, APC876_1WT, APC1450_1WT
P 15	tumour	**TP53273_2**, K_RAS12_1WT, BRAF600_2WT, TP53175_2WT, APC876_1WT, APC1450_1WT
8	normal	K_RAS12_1WT, BRAF600_2WT, TP53175_2WT, APC876_1WT, APC1450_1WT
P 22	tumour	K_RAS12_1WT, BRAF600_2WT, TP53175_2WT, APC876_1WT, APC1450_1WT
9	normal	K_RAS12_1WT, BRAF600_2WT, TP53175_2WT, APC876_1WT, APC1450_1WT
P 14	tumour	K_RAS12_1WT, BRAF600_2WT, TP53175_2WT, APC876_1WT, APC1450_1WT
10	normal	K_RAS12_1WT, BRAF600_2WT, TP53175_2WT, APC876_1WT, APC1450_1WT
P 23	tumour	K_RAS12_1WT, BRAF600_2WT, TP53175_2WT, APC876_1WT, APC1450_1WT
11	normal	K_RAS12_1WT, BRAF600_2WT, TP53175_2WT, APC876_1WT, APC1450_1WT
P 24	tumour	**TP53175_2**, K_RAS12_1WT, BRAF600_2WT, TP53175_2WT, **APC1309_1**, APC876_1WT, APC1450_1WT
12	normal	K_RAS12_1WT, BRAF600_2WT, TP53175_2WT, APC876_1WT, APC1450_1WT
P 25	tumour	**TP53245_1**, K_RAS12_1WT, BRAF600_2WT, TP53175_2WT, **APC1450_1**, APC876_1WT, APC1450_1WT
13	normal	K_RAS12_1WT, BRAF600_2WT, TP53175_2WT, APC876_1WT, APC1450_1WT
P 26	tumour	**BRAF600_2**, K_RAS12_1WT, BRAF600_2WT, TP53175_2WT, **APC1450_1**, APC876_1WT, APC1450_1WT
14	normal	K_RAS12_1WT, BRAF600_2WT, TP53175_2WT, APC876_1WT, APC1450_1WT
E09	tumour	**BRAF600_2**, K_RAS12_1WT, BRAF600_2WT, TP53175_2WT, APC876_1WT, APC1450_1WT
15	normal	K_RAS12_1WT, BRAF600_2WT, TP53175_2WT, APC876_1WT, APC1450_1WT
P 27	tumour	K_RAS12_1WT, BRAF600_2WT, TP53175_2WT, APC876_1WT, APC1450_1WT
16	normal	K_RAS12_1WT, BRAF600_2WT, TP53175_2WT, APC876_1WT, APC1450_1WT
P28	tumour	**TP53273_1**, K_RAS12_1WT, BRAF600_2WT, TP53175_2WT, APC876_1WT, APC1450_1WT
17	normal	K_RAS12_1WT, BRAF600_2WT, TP53175_2WT, APC876_1WT, APC1450_1WT
P11	tumour	K_RAS12_1WT, BRAF600_2WT, TP53175_2WT, **APC1338_1**, APC876_1WT, APC1450_1WT
18	normal	K_RAS12_1WT, BRAF600_2WT, TP53175_2WT, **APC1450_1**, APC876_1WT, APC1450_1WT
P 29	tumour	K_RAS12_1WT, BRAF600_2WT, TP53175_2WT, **APC1450_1**, APC876_1WT, APC1450_1WT
19	normal	**TP53273_2**, K_RAS12_1WT, BRAF600_2WT, TP53175_2WT, APC876_1WT, APC1450_1WT
P 30	tumour	**TP53273_2**, K_RAS12_1WT, BRAF600_2WT, TP53175_2WT, APC876_1WT, APC1450_1WT
20	normal	K_RAS12_1WT, BRAF600_2WT, TP53175_2WT, **APC1450_1**, APC876_1WT, APC1450_1WT
P 31	tumour	K_RAS12_1WT, BRAF600_2WT, TP53175_2WT, APC876_1WT, APC1450_1WT
21	normal	K_RAS12_1WT, BRAF600_2WT, TP53175_2WT, APC876_1WT, APC1450_1WT
E 10	tumour	K_RAS12_1WT, BRAF600_2WT, TP53175_2WT, APC876_1WT, APC1450_1WT
22	normal	K_RAS12_1WT, BRAF600_2WT, TP53175_2WT, APC876_1WT, APC1450_1WT
P 32	tumour	K_RAS12_1WT, BRAF600_2WT, TP53175_2WT, APC876_1WT, APC1450_1WT
23	normal	**TP53273_1**, K_RAS12_1WT, BRAF600_2WT, TP53175_2WT, APC876_1WT, APC1450_1WT
P 33	tumour	**TP53175_2, TP53273_1**, K_RAS12_1WT, BRAF600_2WT, TP53175_2WT, APC876_1WT, APC1450_1WT
24	normal	K_RAS12_1WT, BRAF600_2WT, TP53175_2WT, APC876_1WT, APC1450_1WT
P 34	tumour	K_RAS12_1WT, BRAF600_2WT, TP53175_2WT, APC876_1WT, APC1450_1WT
25	normal	K_RAS12_1WT, BRAF600_2WT, TP53175_2WT, APC876_1WT, APC1450_1WT
P12	tumour	K_RAS12_1WT, BRAF600_2WT, TP53175_2WT, APC876_1WT, APC1450_1WT
26	normal	K_RAS12_1WT, BRAF600_2WT, TP53175_2WT, **APC1378_1, APC1450_1**, APC876_1WT, APC1450_1WT
P 35	tumour	**TP53248_1**, K_RAS12_1WT, BRAF600_2WT, TP53175_2WT, APC876_1WT, APC1450_1WT
27	normal	K_RAS12_1WT, BRAF600_2WT, TP53175_2WT, APC876_1WT, APC1450_1WT
P 36	tumour	K_RAS12_1WT, BRAF600_2WT, TP53175_2WT, **APC1450_1**, APC876_1WT, APC1450_1WT
28	normal	K_RAS12_1WT, BRAF600_2WT, TP53175_2WT, APC876_1WT, APC1450_1WT

 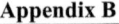

Appendix B

P 37	tumour	K_RAS12_2AS, TP53273_1, K_RAS12_1WT, BRAF600_2WT, TP53175_2WT, APC876_1WT, APC1450_1WT
29	normal	K_RAS12_1WT, BRAF600_2WT, TP53175_2WT, APC876_1WT, APC1450_1WT
P 38	tumour	K_RAS12_1SE, K_RAS12_1WT, BRAF600_2WT, TP53175_2WT, APC876_1WT, APC1450_1WT
30	normal	TP53245_1, K_RAS12_1WT, BRAF600_2WT, TP53175_2WT, APC876_1WT, APC1450_1WT
P 39	tumour	K_RAS12_1WT, BRAF600_2WT, TP53175_2WT, APC876_1WT, APC1450_1WT
31	normal	K_RAS12_1WT, BRAF600_2WT, TP53175_2WT, APC876_1WT, APC1450_1WT
P 40	tumour	K_RAS12_1WT, BRAF600_2WT, TP53175_2WT, **APC1338_1**, APC876_1WT, APC1450_1WT
32	normal	K_RAS12_1WT, BRAF600_2WT, TP53175_2WT, APC876_1WT, APC1450_1WT
P 41	tumour	TP53245_1, K_RAS12_1WT, BRAF600_2WT, TP53175_2WT, APC876_1WT, APC1450_1WT
33	normal	TP53273_1, K_RAS12_1WT, BRAF600_2WT, TP53175_2WT, APC876_1WT, APC1450_1WT
P 42	tumour	**TP53175_2, TP53273_1**, K_RAS12_1WT, BRAF600_2WT, TP53175_2WT, APC876_1WT, APC1450_1WT
34	normal	TP53273_1, K_RAS12_1WT, BRAF600_2WT, TP53175_2WT, APC876_1WT, APC1450_1WT
P 43	tumour	**K_RAS12_2AS**, K_RAS12_1WT, BRAF600_2WT, TP53175_2WT, APC876_1WT, APC1450_1WT
35	normal	K_RAS12_1WT, BRAF600_2WT, TP53175_2WT, APC876_1WT, APC1450_1WT
P 44	tumour	**K_RAS12_2AS**, K_RAS12_1WT, BRAF600_2WT, TP53175_2WT, **APC1309_5del**, APC876_1WT, APC1450_1WT
36	normal	K_RAS12_1WT, BRAF600_2WT, TP53175_2WT, APC876_1WT, APC1450_1WT
P 45	tumour	**K_RAS12_1SE, K_RAS12_2VA**, K_RAS12_1WT, BRAF600_2WT, TP53175_2WT, **APC1554_1ins**, APC876_1WT, APC1450_1WT
37	normal	K_RAS12_1WT, BRAF600_2WT, TP53175_2WT, APC876_1WT, APC1450_1WT
P 46	tumour	**TP53248_1**, K_RAS12_1WT, BRAF600_2WT, TP53175_2WT, APC876_1WT, APC1450_1WT
39	normal	K_RAS12_1WT, BRAF600_2WT, TP53175_2WT, APC876_1WT, APC1450_1WT
P 48	tumour	**APC1450_1**, K_RAS12_1WT, BRAF600_2WT, TP53175_2WT, APC876_1WT, APC1450_1WT
40	normal	K_RAS12_1WT, BRAF600_2WT, TP53175_2WT, APC876_1WT, APC1450_1WT
P 49	tumour	**K_RAS12_2AS, K_RAS12_1SE, TP53273_1**, K_RAS12_1WT, BRAF600_2WT, TP53175_2WT, APC876_1WT, APC1450_1WT
41	normal	**K_RAS13_2AS**, K_RAS12_1WT, BRAF600_2WT, TP53175_2WT, APC876_1WT, APC1450_1WT
P 50	tumour	**K_RAS12_1SE, K_RAS13_2AS, TP53273_1**, K_RAS12_1WT, BRAF600_2WT, TP53175_2WT, APC876_1WT, APC1450_1WT
42	normal	K_RAS12_1WT, BRAF600_2WT, TP53175_2WT, APC876_1WT, APC1450_1WT
P13	tumour	K_RAS12_1WT, BRAF600_2WT, TP53175_2WT, APC876_1WT, APC1450_1WT
43	normal	**K_RAS12_1SE, TP53273_1**, K_RAS12_1WT, BRAF600_2WT, TP53175_2WT, APC876_1WT, APC1450_1WT
P 51	tumour	K_RAS12_1WT, BRAF600_2WT, TP53175_2WT, APC876_1WT, APC1450_1WT
44	normal	**TP53245_1**, K_RAS12_1WT, BRAF600_2WT, TP53175_2WT, **APC1306_1**, APC876_1WT, APC1450_1WT
P 52	tumour	**TP53245_1**, BRAF600_2WT, TP53175_2WT, APC876_1WT, APC1450_1WT
45	normal	K_RAS12_1WT, BRAF600_2WT, TP53175_2WT, APC876_1WT, APC1450_1WT
P 53	tumour	K_RAS12_1WT, BRAF600_2WT, TP53175_2WT, APC876_1WT, APC1450_1WT
46	normal	K_RAS12_1WT, BRAF600_2WT, TP53175_2WT, **APC1554_1ins**, APC876_1WT, APC1450_1WT
P 54	tumour	K_RAS12_1WT, BRAF600_2WT, TP53175_2WT, APC876_1WT, APC1450_1WT
47	normal	K_RAS12_1WT, BRAF600_2WT, TP53175_2WT, APC876_1WT, APC1450_1WT
P 55	tumour	K_RAS12_1WT, BRAF600_2WT, TP53175_2WT, **APC1379_1**, APC876_1WT, APC1450_1WT
48	normal	K_RAS12_1WT, BRAF600_2WT, TP53175_2WT, APC876_1WT, APC1450_1WT
P 56	tumour	K_RAS12_1WT, BRAF600_2WT, TP53175_2WT, **APC1338_1**, APC876_1WT, APC1450_1WT
49	normal	K_RAS12_1WT, BRAF600_2WT, TP53175_2WT, APC876_1WT, APC1450_1WT
P 57	tumour	**K_RAS12_1CY**, K_RAS12_1WT, BRAF600_2WT, TP53175_2WT, APC876_1WT, APC1450_1WT
50	normal	K_RAS12_1WT, BRAF600_2WT, TP53175_2WT, **APC1306_1, APC1554_1ins**, APC876_1WT, APC1450_1WT
P 58	tumour	**TP53175_2**, K_RAS12_1WT, BRAF600_2WT, TP53175_2WT, **APC1367_1**, APC876_1WT, APC1450_1WT
51	normal	K_RAS12_1WT, BRAF600_2WT, TP53175_2WT, APC876_1WT, APC1450_1WT
P 59	tumour	**TP53245_1**, K_RAS12_1WT, BRAF600_2WT, TP53175_2WT, **APC1309_5del**, APC876_1WT, APC1450_1WT
52	normal	K_RAS12_1WT, BRAF600_2WT, TP53175_2WT, APC876_1WT, APC1450_1WT
P 60	tumour	**BRAF600_2, TP53282_1**, K_RAS12_1WT, BRAF600_2WT, TP53175_2WT, APC876_1WT, APC1450_1WT
53	normal	K_RAS12_1WT, BRAF600_2WT, TP53175_2WT, APC876_1WT, APC1450_1WT
P 61	tumour	**BRAF600_2**, K_RAS12_1WT, BRAF600_2WT, TP53175_2WT, APC876_1WT, APC1450_1WT
54	normal	K_RAS12_1WT, BRAF600_2WT, TP53175_2WT, APC876_1WT, APC1450_1WT
P 62	tumour	**TP53175_2**, K_RAS12_1WT, BRAF600_2WT, TP53175_2WT, APC876_1WT, APC1450_1WT
55	normal	K_RAS12_1WT, BRAF600_2WT, TP53175_2WT, APC876_1WT, APC1450_1WT
P 63	tumour	K_RAS12_1WT, BRAF600_2WT, TP53175_2WT, APC876_1WT, APC1450_1WT
56	normal	K_RAS12_1WT, BRAF600_2WT, TP53175_2WT, APC876_1WT, APC1450_1WT
P 64	tumour	K_RAS12_1WT, BRAF600_2WT, TP53175_2WT, APC876_1WT, APC1450_1WT

Appendix C

Lists of all patient samples from the 4 validation cohorts for TFAP2E

Bochum 51 patients analysed with HRM

Extraction No	Patient Number	HRM	HRM II	MEAN	Response	Extraction No	Patient Number	HRM	HRM II	MEAN	Response
Batch One						**Batch Two**					
1	416	U/M	M	M	N	1	81	U	U	U	N
2	344	U/M	U	U	N	2	384	U	U	U	N
3	171	U/M	U	U	N	3	402	U/M	U	U	R
4	49	U	U	U	R	4	173	M	M	M	R
5	354	U	U	U	R	5	467				N
6	202	U/M	U	U	R	6	309	U/M	U	U	R
7	231	U	U	U	N	7	116	U	U	U	R
8	147	U	U	U	N	8	291	U/M	U	U	R
9	429	U	M/U	u	R	9	219	U	U	U	R
10	140	M	M	M	R	10	239	U/M	M	M	N
11	302	U	U	U	N	11	441	U	U	U	N
12	155	M	U	u	R	12	413	M	U	m	N
Batch Three						**Batch Four**					
1	199	M	U	u	R	1	585/3	U/M	M	M	N
2	249	U	U	U	N	2	520/3	U/M	na	u	R
3	247	U/M	U	U	N	3	637/3	U	U	U	N
4	282	U/M	U	U	R	4	W2427/4	U	U	U	N
5	451	U	U	U	N	5	WI836/3	U	M/U	u	R
6	469	U/M	U	u	R	6	W2628/3	U	U	U	R
7	277	U/M	M	M	N	7	WI544/3	U/M	U	U	R
8	303	U/M	U	U	R	8	3007/4	U/M	U	U	N
9	111	M	u	M	N	9	583/3	U/M	M	M	N
10	W2426/4	U	u	U	N	10	2510/4	U	U	U	R
11	W624/3	U	M	u	R	11	W1975/3	M	u	M	N
12	WI 747/3	U/M	U	U	N	12	W2974/4	U/M	U	U	N

Bochum 25 Patients analysed with HRM and Methylight

NR	PMR	ML	HRM	Therapy	Response	NR	PMR	ML	HRM	Therapy	Response
65		-	-	FUFOX	SD						
49	11.95	U	U	FUFOX	PR	172	90.59	M	M	FUFOX	SD
68	8.67	U	M	FUFOX	SD	176	21.73	U	U	CAPOX	PR
81	52.67	M	M	CAPOX	PD	181	7.02	U	U	CAPOX	SD
83	10.74	U	U	FUFOX	PD	184	19.54	U	U	FUFOX	SD
84	34.35	M	M	CAPOX	PR	192	12.17	U	U	FUFOX	PR
100	0.35	U	U	FUFOX	PR	202	56.22	M	M	FUFOX	PR
140	8.36	U	M	CAPOX	SD	214	12.38	U	U	CAPOX	PR
153	12.42	U	M	FUFOX	SD	218	61.37	M	M	FUFOX	PD
155	4.17	U	U	FUFOX	PR	239	5.45	U	M	FUFOX	SD
164	66.35	M	M	CAPOX	SD	247	8.46	U	U	FUFOX	PD
169	5.48	U	U	FUFOX	PR	248	12.96	U	U	FUFOX	PR
171	4.06	U	U	CAPOX	PD	267	29.86	M	U	CAPOX	PR

Appendix C

Mannheim 50 Patients analysed with both Methylight and HRM (all FFPE):

ExtractionNumber	Patient Number	PMR	Methylight	HRM	MEAN	Response
1	307773/07	36.68	M	U	U	R
2	26448/05	65.52	M	M	M	NR
3	8074/06	66.51	M	U	U	R
4	11233/05	13.75	U	U	U	R
5	10766/07	14.07	U	M	M	R
6	13063/06	0.379	U	-	U	R
7	24088/05	0.60	*U*	M	M	NR
8	3968/07	3.282	U	U	U	R
9	8100/05	11.89	U	M	M	NR
10	12438/03	1.609				
11	32598/04	2.442	U	U	U	NR
12	44634/06	3.41	U	U	U	R
13	20888/06		Same as no2			
14	A03936/08	17.16	U	U	U	R
15	11926/04	3.154	U	M	M	NR
16	17778/06	8.616	U	M	M	NR
17	7700/07	12.05	U	U	U	R
18	22159/05	8.244	U	M	M	NR
19	27715/06	10.32	U	M	M	NR
20	19483/03	6.45				
21	19278/05	12.5	U	M/U	U	R
22	2371/06	42.64	M	M	M	NR
23	13721/05	26.18	U	M/U	U	R
24	13280/06	25.96	U	M	M	NR
25	15545/07	20.3	U	m/u	U	R
26	9916/05	35.48	M	U	U	R
27	21274/05	13.66	U	U	U	R
28	1130/07	21.53	U	M	M	NR
29	29443/05	3.256	U	M	M	NR
30	17019/07	30.58	M	M	M	R
31	9867/05	17.85	U	U	U	R
32	29904/05	33.03	M	M	M	R
33	14033/07	26.09	U	U	U	NR
34	22151/06	18.29	U	U	U	R
35	3065/06	4.324	U	U	U	R
36	26942/07	20.63	U	U	U	NR
37	3008/06	19.42	U	M	M	NR
38	19879/07	7.3	U	M	M	NR
39	13293/05	23.12	U	U	U	R
40	6572/08	12.5	U	U	U	R
41	32218/05	10.07	U	U	U	R
42	30813/06	14.8	U	U	U	R
43	6839/07	6.38	U	M	M	NR
44	11961/08	4.321	U	M	M	R
45	52716/07	19.74				
46	25640/03	14.81				
47	31844/06	12.13	U	M	M	R
48	193/05		Same as No31			
49	3562/05	28.84				
50	15078/02	5.977				

Appendix C

Munich 40 Patients analysed with both Methylight and HRM:

Patient Number	PMR	ML	HRM	Response
415	113.20	M	M	NR
416	97.60	M	u	NR
417	83.11	M	m	NR
419	172.15	M	M	NR
420	94.88	M	u	R
421	94.10	M	m	NR
422	94.25	U	U	R
423	29.67	U	u	R
424	12.74	U	U	R
425	11.81	M	M	NR
426	97.52	M	M	NR
427	146.74	M	u	R
428	76.48	U	U	R
429	27.87	U	U	R
430	29.32	U	U	R
431	26.10	M	m	NR
432	33.88	M	M	NR
433	76.59	U	U	R
434	17.08	M	u	NR
435	51.16	U	U	R
436	14.16	M	M	NR
437	63.91	M		NR
438	42.62	M	M	NR
440	55.71	M	M	NR
441	68.15	U	U	R
442	23.82	M		NR
443	83.15	M		NR
444	87.72	M	M	NR
445	140.03	M		NR
446	143.77	M	M	NR
447	244.99	M	M/U	NR
448	74.72	U	U	R
449	5.15	U	U	R
450	14.31	M	M	NR
451	248.66	M		NR
452	56.28	M	m	NR
453	203.29	M	M	NR
454	133.15	M	m	NR
455	137.50	M	m	NR
462	129.88	M	M	NR

Appendix C

Munich additional 30 Patients analysed with MethyLight and HRM:

Extraction Number	Patient Number	PMR	ML	HRM	Response
1	171	25.04	U	U	R
2	194	11.01	U	U	R
3	256	9.64	U	U	NR
4	258	20.33	U	U	R
5	1	2.38	U	U	R
6	52	6.98	U	U	R
7	63	17.95	U	U	NR
9	92	12.13	U	U	R
11	102	45.27	M	M	R
12	123	74.83	M	M	R
13	125	22.13	U	U	R
14	126	6.91	U	U	R
15	133	55.47	M	M	NR
16	136	55.60	M	u	R
17	137	52.99	M	U	R
18	159	56.13	M	M	R
20	3	16.51	U	U	R
21	4	52.77	M	U	R
22	11	24.27	U	U	R
23	18	17.52	U	U	NR
24	19	20.42	U	M	NR
25	20	5.99	U	U	NR
26	24	13.80	U	U	R
27	25	24.40	U	U	r
28	26	13.37	U	U	NR
29	29	46.02	M	U	R
30	4	11.41	U	U	NR
31	5	21.05	U	M	NR

Dresden 45 Patients analysed with MethyLight:

Patient	PMR	HRM	Methylight	Response
D1	58.84		M	NR
D2	11.58		U	R
D4	87.13		M	NR
D5	35.14	m	M	R
D6	258.36		M	NR
D7	28.62		U	R
D8	35.26		M	NR
D9	20.67	u	(U)	Tox
D10	11.32		(U)	-
D11	40.30		M	NR
D12	65.55		M	NR
D13	16.18		U	R
D14	32.76		M	NR
D15	7.51		U	R
D16	58.79	m	M	NR
D17	243.57		M	NR
D18	20.42		U	R
D19	65.97	m	M	NR
D20	12.75		U	R

150

Appendix C

D21	19.60	u	U	R
D22	22.42		U	R
D23	62.28		M	NR
D24	17.34	u	U	R
D25	267.54		M	NR
D26	277.22		M	NR
D27	89.99	m	M	NR
D28	26.30		U	R
D29	16.94	u	U	R
D30	61.77		M	NR
D31	55.55	m	M	NR
D32	28.73	u	(U)	Tox
D33	20.26	u	(U)	Tox
D34	57.00		M	NR
D35	68.52	m	M	NR
D36	34.87	m	M	NR
D37	24.83	u	U	R
D38	81.43	m	M	NR
D39	32.73	m	M	NR
D40	200.08	m	(M)	-
D41	14.44	u	U	R
D42	51.92	m	M	NR
D43	194.86		-	(PR)
D44	9.52		(M)	-
D45	58.84	u/m		

Summary of the results of the patient samples from Bochum (HRM analysis):

	all samples from FFPE material					all patients		
n = 50	R	NR	n = 24	R	NR	both n =74	R	NR
M	2	9	M	1	8	M	3	17
U	23	16	U	10	5	U	33	21
P	0.0374		P	0.013		P	0.0005	

Summary of the results of the patient samples from Mannheim (HRM analysis):

n=42	R	NR	P
M	5	14	0.0001
U	20	3	all FFPE

Summary of the Results of the patients from Munich:

	FFPE material			frozen material			all patients	
n = 36	R	NR	n = 28	R	NR	both n =68	R	NR
M	0	21	M	3	3	M	3	28
U	13	2	U	16	6	U	29	8
P	0.0001		P	ns		P	0.0001	

Summary of the results of the patient samples from Dresden (MethyLight analysis):

n=36	R	NR	P
M	1	22	0.0001
U	13	0	

Appendix D

TESS Predictions for TFAP2A Binding Sites

Gene	Fold change Microarray	No. of putative TFAP2 (alpha/beta) binding sites and conserved (human, mouse, rat, chicken, clawed frog) almost complete (+/-1bp) AP2A consensus sequence (GCCNNNGGC)	Location in the proximal promoter starting with -2000bp of TSS (6-8bp in length) one species vs. conserved	CpG Island
DKK4	-5.32	4 (3) 2	462. 537. **1405.** 1669	no
DKFZP564O0823 = PARM1	-3.20	3 2	624. 854. **1885**	yes
MAGEC2	23.97	10 (6) 0	147. 278. 631. **1303. 1407. 1533.** 1578. **1732.** 1880. 1942	no
ASB4	8.65	6 (2) 1	104. 162. 175. 290. 944. 1406	yes
MUC13	3.12	5 (2) 1	296. 805. **1116. 1358.** 1534	no
PSG5	-2.52	17 (14) 6	5. 45. 267. 344. 441. **563.** 699. 801. 875. 957. **1069.** 1198. **1350.** 1669. 1683. **1716. 1985**	no
ALDH1A3	-2.48	23 (22) 9	176. 370. 378. 466. 527. 668. 721. 851. **1023.** 1230. **1280.** 1362. 1370. 1432. 1622. 1647. 1693. 1726. 1863. 1875. 1970. 1984	yes
PSG1	-2.4	17 (12) 2	24. 323. 420. 497. 541. 640. 677. 778. 852. 934. 1046. 1145. 1196. 1327. 1646. **1693**	no
SYTL3	2.82	2 0	1046. 1268	
SLC1A2	2.44	17 (15) 4	465. 823. **1027.** 1117. **1278. 1331. 1344. 1441. 1454. 1519.** 1621. 1697. 1725. 1851. 1871. 1912. 1957	yes
VIL1	2.43	8 1	71. 562. 1366. **1473. 1499.** 1827. 1900. 1968	no
GMFG	2.39	5 (3) 1	530 836 1315 1322 1698	no
CKMT2	2.36	11 (8) 2	538 724 **1030** 1405 1441 1582 1684 1780 **1802** 1866 1954	no
GABRA3	-2.34	8 (3) 0	94 861 1164 1241 **1414** 1524 1930 1943	yes
RBMY2EP	-2.29	5 (3) 1	142 189 242 825 **1728**	yes
CSGALNACT1	-2.24	10 (5) 3	60 75 165 223 235 772 997 1160 **1432** 1705	yes
PSG7	-2.23	12 (10) 2	204 281 317 380 501 600 813 889 1006 1287 1606 **1955**	no
BNC2	-2.15	19 (18) 0	754 763 781 926 1124 1156 1214 1223 1235 1258 1398 1482 1543 **1605** 1638 1789 1815 1882 1973	yes
NEDD9	-2.14	3 0	132 1102 1156	no

AUC Blood and Polyps

GENE	AUC Polyp vs. Blood	AUC Polyp vs. Normal
TFAP2E	0.91	0.65
TUSC3	0.96	**0.94**

Appendix E
Vectors

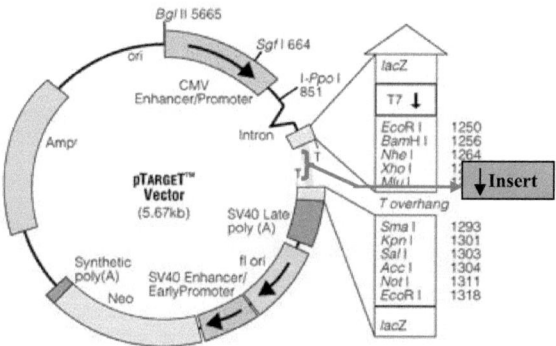

pTARGET vector (Promega), which was prepared by digestion with EcoRV followed by addition of a 3´-T overhang to each end. For preparation of empty vector controls, the overhangs were filled with a deoxyadenine by a Taq polymerase and ligated to recirculate the vector. To linearize the vector again, it was cut with SMAI and thymidines added by a Taq polymerase.

Inserts: TFAP2E coding sequence (bp 220-1550 Genbank Acc. No. NM_178548.2)

```
 200 accgccccat gctggtgcac acctactccg ccatggagcg
 241 ccccgacggg ctgggagcag ctgccggcgg ggcccgcctg tcgtctctgc cccaggcggc
 301 ctacgggccg gcgcccccgc tctgccacac gccggccgcc acagctgccg ccgaattcca
 361 gccgccctac ttcccgccgc cctacccgca gccaccgctg ccctacggtc aggcgcccga
 421 cgccgccgca gcctttcccc acctggcagg ggaccatat ggcggcctgg cgccccctggc
 481 gcagccgcag cctcctcagg ccgcctgggc cgcgccccgc gcagccgccc gcgcccacga
 541 ggagcctccc ggcctgctgg caccgcccgc ccgcgccctg ggccttgacc cgcgccgtga
 601 ctatgccact gccgtgcccc ggctcctgca cggcctggcc gacggcgcgc acggcctggc
 661 agacgcacct ctcggccttc cggggctggc ggccgccccc ggtctggagg acctgcaggc
 721 aatggacgag ccgggaatga gcctcctaga ccagtccgtg atcaagaaag tgcccatccc
 781 ctccaaagcc agcagcctct cagccctctc cttggccaaa gacagcctgg tgggcggcat
 841 cacaaatcct ggtgaggtct tctgctccgt gcccggccgg cttctcactgc tcagctcaac
 901 gtccaagtac aaggtgacgg tggggaggt gcagcggcga ctctcgcctc ccgagtgcct
 961 caacgcctcc ctcctggggg gtgtcctccg cagggcaag tccaaaaatg ggggccggtg
1021 tttgcgggaa cggttagaga agattgggct caacctgcca gctggccgtc gcaaggccgc
1081 caatgtgacg ctgctgactt cgctagtgga aggagaggcc gtgcacctgg cccgagactt
1141 cggttacgtc tgtgagacgg agttcccagc caaggcagct gccgagtacc tgtgccgaca
1201 gcacgctgac ccgggggagc tgcacagccg caagagcatg ctgctggctg ccaagcagat
1261 ctgcaaggag tttgcagact tgatggctca ggaccgctca ccgctgggca acagccgccc
1321 agcactcatc ctggagcccg gagtacagag ctgcttgaca cactttagcc tcatcaccca
1381 tggcttcggt gggcctgcca tctgtgctgc cctcactgcc ttccagaact atttgctgga
1441 gtcactcaag gggctggaca agatgtttct aagcagtgtg ggcagtgggc atggtgaaac
1501 caaggcttcg gagaaggatg ccaagcatcg gaataactg cttctcccac cccatccct
```

Appendix E

RXFP3 coding sequence (bp 348-1780 Genbank Acc. No. NM_016568.3)

```
                        340 tagaggtacc tgcgcatgca
 361 gatggccgat gcagccacga tagccaccat gaataaggca gcaggcgggg acaagctagc
 421 agaactcttc agtctggtcc cggaccttct ggaggcggcc aacacgagtg gtaacgcgtc
 481 gctgcagctt ccggacttgt ggtgggagct ggggctggag ttgccggacg gcgcgccgcc
 541 aggacatccc ccgggcagcg gcggggcaga gagcgcggac acagaggccc gggtgcggat
 601 tctcatcagc gtggtgtact gggtggtgtg cgccctgggg ttggcgggca acctgctggt
 661 tctctacctg atgaagagca tgcagggctg gcgcaagtcc tctatcaacc tcttcgtcac
 721 caacctggcg ctgacggact ttcagtttgt gctcaccctg cccttctggg cggtggagaa
 781 cgctcttgac ttcaaatggc ccttcggcaa ggccatgtgt aagatcgtgt ccatggtgac
 841 gtccatgaac atgtacgcca gcgtgttctt cctcactgcc atgagtgtga cgcgctacca
 901 ttcggtggcc tcggctctga agagccaccg gacccgagga cacggccggg gcgactgctg
 961 cggccggagc ctggggggaca gctgctgctt ctcggccaag tgcgctgtgtg tgtggatctg
1021 ggcttttggcc gcgctggcct cgctgcccag tgccattttc tccaccacgg tcaaggtgat
1081 gggcgaggag ctgtgcctgg tgcgtttccc ggacaagttg ctgggccgcg acaggcagtt
1141 ctggctgggc ctctaccact cgcagaaggt gctgctgggc ttcgtgctgc cgctgggcat
1201 cattatcttg tgctacctgc tgctggtgcg cttcatcgcc gaccgccgcg cggcggggac
1261 caaaggaggg gccgcggtag ccggaggacg cccgaccgga gccagcgcc ggagactgtc
1321 gaaggtcacc aaatcagtga ccatcgttgt cctgtccttc ttcctgtgtt ggctgcccaa
1381 ccaggcgctc accacctgga gcatcctcat caagttcaac gcggtgccct tcagccagga
1441 gtatttcctg tgccaggtat acgcgttccc tgtgagcgtg tgcctagcgc actccaacag
1501 ctgcctcaac ccgtcctct actgcctcgt gcgccgcgag ttccgcaagg cgctcaagag
1561 cctgctgtgg cgcatcgcgt ctccttcgat caccagcatg cgcccccttca ccgccactac
1621 caagccggag cacgaggatc aggggctgca ggccccggcg ccgccccacg cggccgcgga
1681 gccggacctg ctctactacc cacctggcgt cgtggtctac agcgggggc gctacgacct
                        1741 gctgcccagc agctctgcct actgacgcag gcctcaggcc cagggcgcgc cgtcggggca
```

TUSC3 coding sequence isoform A (bp 324-1423 Genbank Acc. No. NM_006765.3)

```
                        320 tggaggagac actgccctgc cgcgatgggg gccggggcg
 361 ctccttcacg ccgtaggcaa gcggggcggc ggctgcggta cctgcccacc gggagctttc
 421 ccttccttct cctgctgctg ctgctctgca tccagctcgg gggaggacag aagaaaaagg
 481 agaatctttc agctgaaaaa gtagagcagc tgatggaatg gagttccaga cgctcaatct
 541 tccgaatgaa tggtgataaa ttccgaaaat ttataaaggc accacctcga aactattcca
 601 tgattgttat gttcactgct cttcagcctc agcggcaggtg ttctgtgtgc aggcaagcta
 661 atgaagaata tcaaatactg gcgaactcct ggcgctattc atctgctttt tgtaacaagc
 721 tcttcttcag tatggtggac tatgatgagg ggacagacgt ttttcagcag ctcaacatga
 781 actctgctcc tacattcatg catttcctc caaaaggcag acctaagaga gctgatactt
 841 ttgacctcca aagaattgga tttgcagctg agcaactagc aaagtggatt gctgacagaa
 901 cggatgttca tattcgggtt ttcagaccac ccaactactc tggtaccatt gcttttggccc
 961 tgttagtgtc gcttgttgga ggtttgcttt atttgagaag gaacaacttg gagttcatct
1021 ataacaagac tggttgggcc atggtgtctc tgtgtatagt ctttgctatg acttctggcc
1081 agatgtggaa ccatatccgt ggacctccat atgctcataa gaacccacac aatggacaag
1141 tgagctacat tcatcgggagc agccaggctc agtttgtggc agaatcacac attattctgg
1201 tactgaatgc cgctcatcacc atggggatgg ttcttctaaa tgaagcagca acttcgaaag
1261 gcgatgttgg aaaaagacgg ataatttgcc tagtgggatt gggcctggtg gtcttcttct
1321 tcagtttctt actttcaata tttcgttcca agtaccacgg ctatccttat agtgatctgg
1381 actttgagtg agaagatgtg atttggacca tggcacttaa aaactctata acctcagctt
```

Appendix E

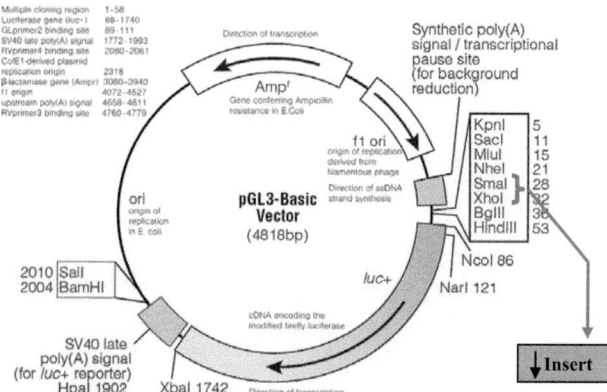

The pGL3-Basic Luciferase vector (Promega), which contains a modified coding region for firefly (Photinus pyralis) luciferase. For linearization of the vector, it was cut with SMAI and thymidines added by a Taq polymerase.

Inserts: DKK4 Promoter (-2000 bp from TSS) Genbank Acc. No. NC_000008, AP2 binding sites are underlined

```
   1 gcctcagcct cccgagtagc tgggattaca ggtgaccacc accacgcccg actagttttt
  61 atattttag tagagacagg gtttcaccat attggccagg ctggtctcga actcctgacc
 121 tcaggtgatc catccaccct agcctcccaa agtggtgaga ttacaggcgt cagccaccgc
 181 acgcggccta cttttttttt tttttttttt tttgtggaga caggatctca atatgttgcc
 241 caggctcgtc ttgaactcct gagctcagga gatcttccca ccttggcctc ccaaagtgct
 301 gggattacag gtgtgagcca ctgtgcccgg ccattggagt ttcaaatttt attcctaaca
 361 tgtaatccga ggctctgttc catgccaagt gctctaagtg ctcagtggct gttgtgccg
 421 tgttacagga gttcccgtca gttctttcgc ctgtgtatat tgccaggggc ctttggtaac
 481 ctgaaattcc cttacagtga ggcagcagga agcagagaaa cccggcagag acagaaccct
 541 gggggctctt tcctttatcc ttggtcataa ttgtacaact tgaaaataaa ttctgtaaca
 601 cccctagtca ccagggttat catgctatca aagggtgttg tttgatgtca gtttaaggga
 661 ttaaaacaag taacttccat acgagaggta gacaagcaat accaacccaa ggacaagggc
 721 caagtagtag gaaggctcga atcacctgca tcaccctcgg aatggcactg cattactcaa
 781 aattgagaat gttcccatgc acagaatgct gagattcatg tcaatgaatt gatgttaggt
 841 tgttctgcag gaaaatgtag taaataataa gattatcatt tatggcatag aaaaaatgca
 901 ttttagatga ctctatgcta agagcaatga aacaatgaag agaaaaatg atttttgat
 961 ccagatctag attgttctac ccttaaagta tttaagacag tgataagcac atgtctaaat
1021 tacctgacta atagtccaaa tggaaaaaaa tcagacagga aagggatgaa gcagaagttt
1081 taaaaaattc gtattaggaa aattggaaat acatgaaatt tcatctgaaa tcagccagag
1141 acttggatcc ttgttctgct taagtgagtt ttacataatg caacacaaat taaatatgtt
1201 gctgataaaa tggacccaaa gcttcgtact gaattgtgta aaatgcgcta tcattgttcc
1261 caccattgaa aaatgggatt ttcataaatt acattatttt cctaataacc ctaacttcaa
1321 acagttcaat ttttcattta aggaaagtaa tttcccaaat ggcatttagt cagaacaggt
1381 gctgttacgt tcacaactt taagcggttg ggattttgac atctggatgc cagagctaat
1441 caggggcgt ttttctcttt tctatctgtc ctcactatct gaagcccagt ttacagcttc
1501 tcagtgtatt ttgtacctca ttagtttttg gcatcctcta tagcatattg gccagtatga
1561 ttcatccttt ctttgagcac aattagctgt gaaaacagat cttagggcag actccctccc
1621 tacccccaaa ggattactga aggtaggaaa tgcaggtgat tatcagagtt tgcctttgat
1681 acagacatcc tgctctgccg tggccctttg aactaacttt gatatgcaat aattaagagg
1741 gattgaaccc ctggaggaga agccgcatgg ccctgcctct tctctcttc tattgagtcc
1801 ttgttttgaa ctattgatca aacagatttg aagggatttg ttgaagcctg tgtggggcag
1861 gaggaaggaa cggagcgggg aggaggcaca tctggttata aatagtttca ggaggaacct
1921 gctggtcaga cttgctcag ccgatttcac gcaccttact cagactaagg tttacttctt
1981 tcagaaaaag cagtgacaaa cagacgacgt gctgagctgc cagcttagtg aagctctgc
```

Appendix E

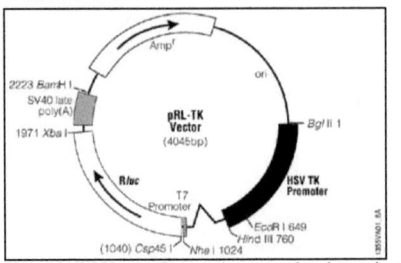

The pRL-TK Vector (Promega) was used as internal control reporter as intended in combination with the pGL3 vector and cotransfected in CRC cells. It contains a cDNA (Rluc) encoding Renilla luciferase, which was originally cloned from the marine organism Renilla reniformis (sea pansy).

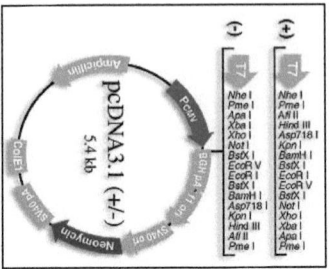

The pcDNA3.1 vector (Invitrogen) was digested with the restriction enzymes BamHI and NotI. Inserts:

DKK4 coding sequence (bp 112-786 Genbank Acc. No. NM_014420.2)

```
  1 cagacgacgt gctgagctgc cagcttagtg gaagctctgc tctgggtgga gagcagcctc
 61 gctttggtga cgcacagtgc tgggaccctc caggagcccc gggattgaag gatggtggcg
121 gccgtcctgc tggggctgag ctggctctgc tctccccctgg gagctctggt cctggacttc
181 aacaacatca ggagctctgc tgacctgcat ggggcccgga agggctcaca gtgcctgtct
241 gacacggact gcaataccag aaagttctgc ctccagcccc gcgatgagaa gccgttctgt
301 gctacatgtc gtgggttgcg gaggaggtgc cagcgagatg ccatgtgctg ccctgggaca
361 ctctgtgtga acgatgtttg tactacgatg gaagatgcaa cccaatatt agaaaggcag
421 cttgatgagc aagatggcac acatgcagaa ggaacaactg ggcacccagt ccaggaaaac
481 caacccaaaa ggaagccaag tattaagaaa tcacaaggca ggaagggaca agagggagaa
541 agttgtctga gaactttttga ctgtggccct ggacttttgct gtgctcgtca tttttggacg
601 aaaatttgta agccagtcct tttggaggga caggtctgct ccagaagagg gcataaagac
661 actgctcaag ctccagaaat cttccagcgt tgcgactgtg gccctggact actgtgtcga
721 agccaattga ccagcaatcg gcagcatgct cgattaagag tatgccaaaa aatagaaaag
781 ctataaatat ttcaaaataa agaagaatcc acattgcaaa aaaaaaaaaa aaaaa
```

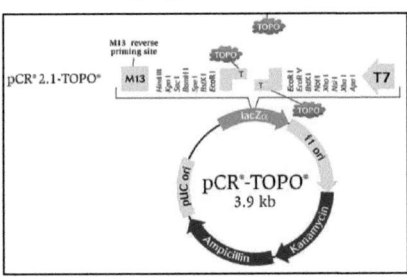

The pCR 2.1 TOPO vector, which was used for subcloning of PCR products and contains 3´-T overhangs for direct ligation of Taq-amplified PCR products with topoisomerase.

Appendix F

List of used Chemicals and Equipment

SOLUTIONS, CHEMICALS, KITS, CELLS, ETC.	MANUFACTURER
0.5% Trypsin EDTA 1X	Invitrogen, Carlsbad, USA and Darmstadt, Germany
1X TAE buffer	selfmade, pulvers from Sigma Aldrich
2-Mercaptoethanol	Sigma Aldrich Co., Steinheim, Germany
2-Propanol, pro analysi	Merck KGaA, Darmstadt, Germany
Agarose powder "Biozym LE Agarose"	Biozym Scientific GmbH
CpGenome Universal Methylated DNA	CHEMICON, Temecula, USA and Milipore, Schwalbach, Germany
DMEM Medium	Invitrogen
Ethanol, absolut pro analysi	Merck KGaA, Darmstadt, Germany
Ethidium bromide solution 10mg/ml	Sigma Aldrich Co., Steinheim, Germany
Foetal bovine serum	Invitrogen
GeneRuler 50bp DNA Ladder	Fermentas GmbH, St. Leon-Rot, Germany
Geneticrin 50mg/ml „Active Geneticrin"	Invitrogen
GoTaq Green Mastermix	Promega GmbH, Mannheim, Germany
L-Glutamine 200mM, 100X	Invitrogen
LightCycler480 Probes Master	Roche, Penzberg, Germany
Lipofectamin 2000	Invitrogen
Opti-MEM I Medium	Invitrogen
Opti-MEM I Reduced Serum Medium	Invitrogen
PBS pH 7,4 1X and DPBS	Invitrogen
Penicillin/Streptomycin „Pen Strep" (10.000U/ml Penicillin; 10.000 µg/ml Streptomycin)	Invitrogen
RNase ZAP	Sigma Aldrich Co., Steinheim, Germany
5-Azacytidine (10µM)	Sigma Aldrich Co., Steinheim, Germany
Dimethyl sulfoxide, minimum 99,5% GC	Sigma Aldrich Co., Steinheim, Germany
Water, HPLC Gradient Grade	Roth GmbH, Karlsruhe, Germany
EpiTect Control DNA, methylated	Qiagen, Hilden, Germany
Universal Methylated Human DNA Standard	Zymo Research Europe, Freiburg, Germany
MethylCode Bisulfite Conversion Kit	Invitrogen
QIAamp DNA Mini Kit	Qiagen, Hilden, Germany
RNase-Free DNase Set	Qiagen
RNeasy Mini Kit and QIAShredder	Qiagen
Verso cDNA Kit	Thermo Fisher Scientific, Darmstadt, Germany
MethylCode Bisulfite Conversion Kit	Invitrogen
QIAamp DNA Mini Kit	Qiagen
RNase-Free DNase Set (50)	Qiagen
DNeasy Blood and Tissue Kit	Qiagen
QIAamp DNA FFPE Tissue Kit	Qiagen
QIAamp DNA Micro Kit	Qiagen
RNeasy Mini Kit	Qiagen
EZ DNA Methylation and EZ DNA Methylation Gold	Zymo Research Europe
EpiTect Bisulfite Kit	Qiagen
Imprint DNA Modification Kit	Sigma
M.SssI methyltransferase and SAM	New England Biolabs GmbH, Frankfurt am Main, Germany
Digestion enzymes BamHI, SMAI, NotI, EcoRV	Promega, Fermentas, New England Biolabs
T4 DNA Ligase	Promega, Fermentas, New England Biolabs
EpiTect MethyLight PCR Kit	Qiagen
PureLink Quick Gel Extraction and PCR Purification Combo Kit	Invitrogen
QuantiTect Probe PCR Kit	Qiagen

Appendix F

EpiTect HRM PCR Kit	Qiagen
LightCycler 480 High Resolution Melting Master	Roche
MeltDoctor HRM Master Mix	Invitrogen
LightCycler 480 SYBR Green Master	Roche
Jump Start Red Taq Ready Mix	Sigma
Qiaquick PCR purification kit	Qiagen
pTARGET, pGL3-basic and pRLTK vectors	Promega
pcDNA 3.1 vector	Invitrogen
JM109 competent E.coli cells	Promega
DH5 alpha Competent Cells	Invitrogen
Top 10 One Shot Competent Cells	Invitrogen
Satisfection Transfection Reagent	Agilent Technologies, Waldbronn, Germany
anti-FLAG antibody (polyclonal rabbit)	Sigma
Human Gene 1.0 ST Expression Array	Affymetrix, High Wycombe, United Kingdom
MTT reagent	Sigma
Cell Proliferation ELISA, BrdU Kit	Roche
Collagen R (rat)	SERVA Electrophoresis GmbH, Heidelberg, Germany
recombinant Relaxin 3 hormone	Phoenix Europe GmbH, Karlsruhe Germany
recombinant Tumor Necrosis Factor alpha protein	Roche and Abcam plc.
AP2epsilon N-12 antibody (goat polyclonal)	Santa Cruz Biotechnology Inc., Heidelberg, Germany
TFAP2E antibody (rabbit polyconal)	Abcam plc. Cambridge, United Kingdom
TUSC3 antibody (rabbit polyclonal)	Abcam
Trichostatin A	Sigma
SDS	Carl Roth
TEMED	Carl Roth
Rotiphorese Gel (polyacrylamid)	Carl Roth GmbH, Karlsruhe, Germany
Paraformaldehyde 4%	Roth
BSA	Carl Roth
Tween 20	Carl Roth
Triton X-100	Carl Roth
Milk powder	Carl Roth
ECL Solution (ready to use HRP-Substrat)	GE Healthcare Limited, Buckinghamshire, UK
Methanol	Carl Roth
HRP goat anti-rabbit and donkey anti-goat polyclonal secondary antibodies	Invitrogen
Alexa Flour 488 goat anti-rabbit secondary antibody	Invitrogen

EUQIPMENT, CONSUMABLES	TYPE	MANUFACTURER
Balance	BL130	Sartorius AG, Göttingen, Germany
Centrifuge	Centrifuge 5810 R	Eppendorf, Hamburg, Germany
Microcentrifuge	Centrifuge 5415 D	Eppendorf, Hamburg, Germany
Microwave	not specified	Siemens, Munich, Germany
Power Supply for gel electrophoresis system	PowerPac Basic	Bio-Rad Laboratories GmbH, Munich, Germany
Real-Time PCR Analysis Software	Light Cycler®480 Software 1.5	Roche, Mannheim, Germany
Real-Time PCR Machine	LightCycler®480 Instrument	Roche, Mannheim, Germany
Rotor-Stator Homogenizer	DIAX 900	Heidolph, Schwabach, Germany
Thermocycler	GeneAmp® PCR System 9700	Applied Biosystems, Foster City, USA
Thermomixer	Thermomixer compact	Eppendorf, Hamburg, Germany
UV-Illumination System	Universal Hood II	Bio-Rad Laboratories GmbH, Munich, Germany
Vortexmachine	REAX top	Heidolph, Schwabach, Germany
Detection Software for agarose gels	QuantityOne 4.5.2	Bio-Rad Laboratories GmbH, Munich, Germany

Appendix F

Gel electrophoresis system with - small gel chamber - large gel chamber and appropriate combs	- MINI-SUB® CELL GT - WIDE MINI-SUB® CELL GT	Bio-Rad Laboratories GmbH, Munich, Germany
Incubator for cell culture	HERA CELL 240	Heraeus Holding GmbH, Hanau, Germany
Microscope for cell culture	Axiovert 40CFL	Carl Zeiss AG, Oberkochen, Germany
Variable pipettes	-	Eppendorf, Hamburg, Germany
1,5ml microcentrifuge tubes „Microtube 1,5ml Safe Seal"	-	Sarstedt AG &Co., Nümbrecht, Germany
2,0ml microcentrifuge tubes	-	Eppendorf, Hamburg, Germany
200µl PCR Tubes "Multiply®-µStrip Pro 8-strip"	-	Sarstedt AG &Co., Nümbrecht, Germany
50ml falcons	-	BECTON DICKINSON, Heidelberg, Germany
96 well plates „Light Cycler®480 Multiwell Plate 96"	-	Roche, Mannheim, Germany
Sealing Foil "Light Cycler®480 Sealing Foil"	-	Roche, Mannheim, Germany
150mm tissue culture dish	-	BECTON DICKINSON, Heidelberg, Germany
24-well tissue culture plates	-	BECTON DICKINSON, Heidelberg, Germany
$25cm^2$ and $75cm^2$ tissue culture flasks	-	BECTON DICKINSON, Heidelberg, Germany
96-well tissue culture plates	-	BECTON DICKINSON, Heidelberg, Germany
Medium filtration system "Stericup®"	-	Millipore Corportion, Billerica, USA
Transwell Matrigel Chambers	-	BD Biosciences, Heidelberg, Germany
BD Falcon BioCoat Culture Slides	-	BD Biosciences

i want morebooks!

Buy your books fast and straightforward online - at one of world's fastest growing online book stores! Environmentally sound due to Print-on-Demand technologies.

Buy your books online at
www.get-morebooks.com

Kaufen Sie Ihre Bücher schnell und unkompliziert online – auf einer der am schnellsten wachsenden Buchhandelsplattformen weltweit! Dank Print-On-Demand umwelt- und ressourcenschonend produziert.

Bücher schneller online kaufen
www.morebooks.de

VDM Verlagsservicegesellschaft mbH
Heinrich-Böcking-Str. 6-8   Telefon: +49 681 3720 174   info@vdm-vsg.de
D - 66121 Saarbrücken   Telefax: +49 681 3720 1749   www.vdm-vsg.de

Printed by Books on Demand GmbH, Norderstedt / Germany